Erratum slip

Eighth European Congress on Fluid Machinery for the Oil, Gas, and Petrochemical Industry
IMechE Conference Transactions 2003–1
ISSN 1356–1448
ISBN 1 86058 384 9

Attached are figures to be included in the following paper – pages 79–88.

C603/023/2003

Ultra-high pressure seawater injection pumps

B GERMAINE
Sulzer Pumps UK Limited

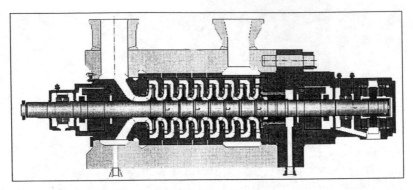

Figure 1   Traditional in-line "stacked impeller" arrangement

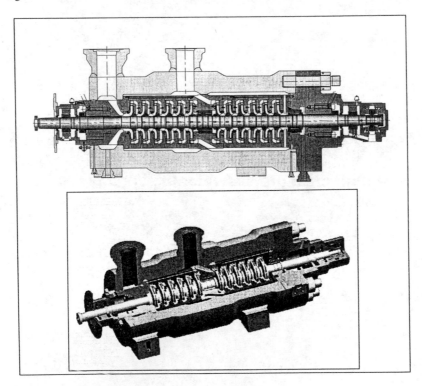

Figure 2   Back to back "opposed impeller" arrangement

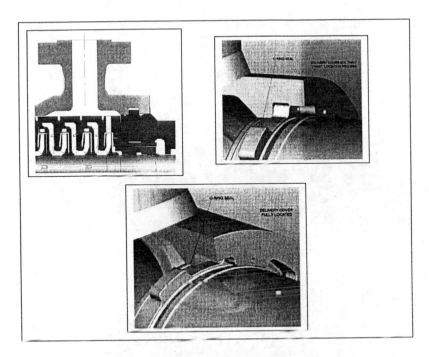

Figure 3   Sulzer patented "Twist-lock" cartridge retaining feature

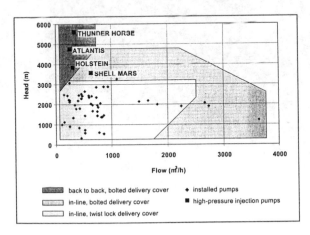

Figure 4   Injection pump range chart showing with deepwater injection pumps highlighted

Figure 5   Thunder Horse pump casing showing discharge cover bolting

Figure 6   Shell Mars pump casing showing very large section thickness at suction end

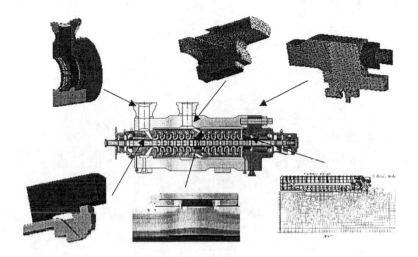

Figure 7   Finite Element Analyses (FEA) that were undertaken on the Thunder Horse
Project to verify the design

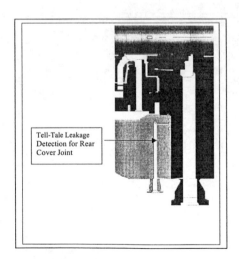

Tell-Tale Leakage
Detection for Rear
Cover Joint

Figure 8   "Tell-Tale" leakage detection design applied to discharge cover primary
face seal

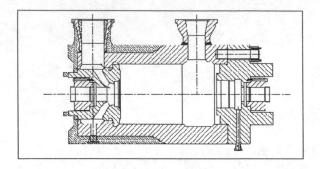

Figure 9   Barrel design showing material requirements for both low and high pressure
suction chamber designs

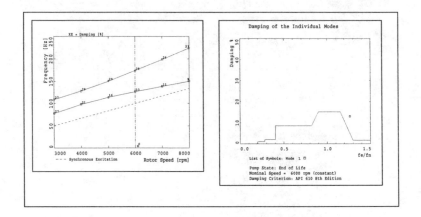

Figure 10   Campbell and Damping Diagrams in accordance with API 610 8th Edition
calculated for the Thunder Horse injection pump. Note that the Damping
plot is for end of life condition.

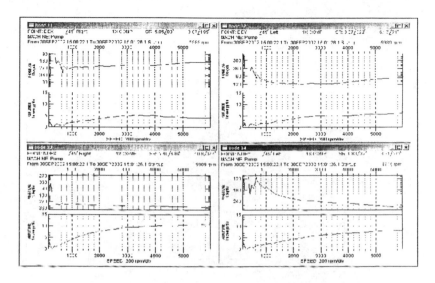

Figure 11   Bode (run-down) plots taken during Thunder Horse prototype tests

Figure 12   Tungsten carbide arrangement for balance piston and balance drum

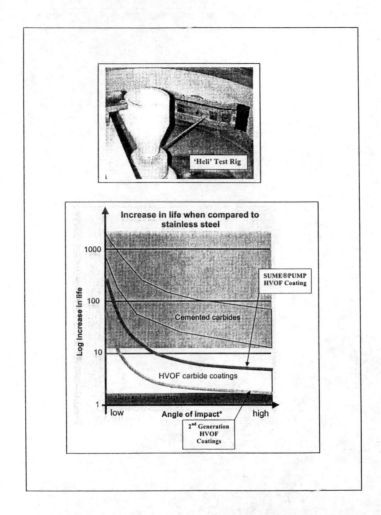

Figure 13   Summary of material test data obtained from both 2-body and "Heli" test rigs

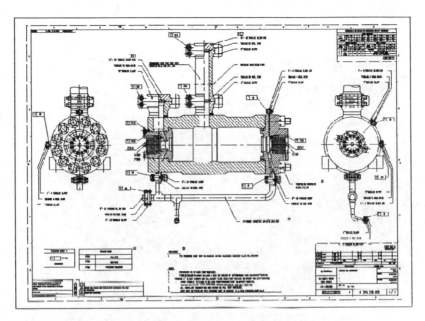

Figure 14   Pressure test arrangement for the Thunder Horse injection pump casing, discharge cover/bolting, suction casing and balance return line

Figure 15   Prototype testing of the Thunder Horse injection pump on the Leeds test stand

Figure 16   The Thunder Horse pump package during string testing, Leeds test facility

Fluid Machinery for the Oil, Gas, and Petrochemical Industry

Conference Organizing Committee

H Carrick (Chairman)
DuPont, UK

I Rhodes
Cranfield University, UK

P Ward
Foster Wheeler Energy Limited, UK

A Grant
Weir Pumps Limited, UK

D Redpath
BP Amoco Oil, UK

R Bouman
APTS, The Netherlands

IMechE
Conference Transactions

I MECH E

Eighth European Congress on

Fluid Machinery for the Oil, Gas, and Petrochemical Industry

31 October–1 November 2002
Bilderberg Europa Hotel, The Hague, The Netherlands

Organized by
The Fluid Machinery Group of the Institution of
Mechanical Engineers (IMechE)

Sponsored by
MAN TURBO

Co-sponsored by
Koninkliike Vlaamse Ingenieursvaremiging
VDI–Gesellschaft Energietechnik
The Engineering Institiute of Canada
CSME
BPMA
EEMUA
KIVI

IMechE Conference Transactions 2003–1

**Professional
Engineering
Publishing**

Published by Professional Engineering Publishing Limited for The Institution of
Mechanical Engineers, Bury St Edmunds and London, UK.

ISSN 1356–1448
ISBN 1 86058 384 9

A CIP catalogue record for this book is available from the British Library.

Printed by The Cromwell Press, Trowbridge, Wiltshire, UK

Related Titles of Interest

Title	Editor/Author	ISBN
Advances of CFD in Fluid Machinery Design	R L Elder, A Tourlidakis, and M K Yates	1 86058 353 9
Composites for the Offshore Oil and Gas Industry	IMechE Seminar	1 86058 229 X
Fluid Film Bearings – Recent Advances in Design and Performance	IMechE Seminar	1 86058 278 8
Fluid Machinery for the Oil, Petrochemical, and Related Industries	IMechE Conference	1 86058 217 6
Handbook of Mechanical Works Inspection	C Matthews	1 86058 047 5
How Did That Happen? – Engineering Safety and Reliability	W Wong	1 86058 359 8
Practical Guide to Engineering Failure Investigation	C Matthews	1 86058 086 6
Process Fan and Compressor Selection	J Davidson and O von Bertele	0 85298 825 7
Reliability of Sealing Systems for Rotating Machinery	IMechE Seminar	1 86058 245 1
Subsea Control and Data Acquisition	L Adriaanse, J H Neuenkirchen, J Cattanach, S Moe, C Eriksen, and H Clayton	1 86058 257 5

For the full range of titles published by Professional Engineering Publishing contact:

Marketing Department
Professional Engineering Publishing Limited
Northgate Avenue
Bury St Edmunds
Suffolk
IP32 6BW
UK

Tel: +44 (0) 1284 763277
Fax: +44 (0) 1284 718692
E-mail: marketing@pepublishing.com
www.pepublishing.com

Contents

Safety and Environment

Operations and Maintenance

Retrofits and Upgrades

Additional Papers

Fluid Machinery Design

C603/001/2003

Optimization of screw compressor design

N STOSIC, I K SMITH, and **A KOVACEVIC**
Centre for Positive Displacement Compressor Technology, City University, London, UK

SYNOPSIS

Ever increasing demands for efficient screw compressors require that compressor designs are tailored upon their duty, capacity and manufacturing capability. A suitable procedure for optimisation of screw compressor shape, dimension and operating parameters is described here, which results in the most appropriate design for a given compressor duty. It is based on a rack generation algorithm for rotor profile combined with a numerical model of the compressor fluid flow and thermodynamic processes. Compressors thus designed achieve higher delivery rates and better efficiencies than those using traditional approaches. Some optimization issues of the rotor profile and compressor parts are discussed, using a 5/6-106 mm screw compressor to illustrate the results. It is shown that the optimum rotor profile, compressor speed, oil flow rate and temperature may significantly differ when compressing different gases or vapours.

Key Words: Screw compressor design, rotor lobe optimal profiling, numerical modelling

1. INTRODUCTION

The screw compressor is a positive displacement rotary machine. It consists essentially of a pair of meshing helical lobed rotors, which rotate within a fixed casing that totally encloses them, as shown in Fig 1. The space between any two successive lobes of each rotor and its surrounding casing forms a separate working chamber of fixed cross sectional area. The length of this chamber varies as rotation proceeds due to displacement of the line of contact between the two rotors. It is a maximum when the entire length between the lobes is unobstructed by meshing contact with the other rotor. It has a minimum value of zero when there is full meshing contact with the second rotor at the end face. The two meshing rotors effectively form a pair of helical gear wheels with their lobes acting as teeth.

As shown in right top side of the figure, gas or vapour enters from the front and on top, through an opening, mainly in the front plane of the casing which forms the low pressure or inlet port. It thus fills the spaces between the lobes, starting from the ends corresponding to A and C in the lightly shaded area. As may be seen, the trapped volume in each chamber increases as rotation proceeds and the contact line between the rotors recedes. At the point where the maximum volume is filled, the inlet port terminates and rotation proceeds without any further fluid admission in the region corresponding to the darkly shaded area.

Fig. 1 Screw compressor principal mechanical parts

Viewed from the bottom, it may be seen that the darkly shaded area begins, from the end corresponding to A and C, at the point where the male and female rotor lobes start to reengage on the underside. Thus, from that position, further rotation reduces the volume of gas or vapour trapped between the lobes and the casing. This causes the pressure to rise. At the position where the trapped volume is sufficiently reduced to achieve the required pressure rise, the ends of the rotors corresponding to D and B are exposed to an opening on the underside of the casing, which forms the high pressure or discharge port. This corresponds to the lightly shaded area at the rear end. Further rotation reduces the trapped volume causing the fluid to flow out through the high pressure port at approximately constant pressure. This continues until the trapped volume is reduced to virtually zero and all the gas trapped between the lobes at the end of the suction process, is expelled. The process is then repeated for each chamber. Thus there is a succession of suction, compression and discharge processes

   C603/001/2003 © With Author 2003

achieved in each rotation, dependent on the number of lobes in the male and female rotors. If the direction of rotation of the rotors is reversed, fluid will flow in to the machine at the high pressure end and out at the low pressure end and it will act as an expander.

Screw compressor rotors of various profiles can be flexibly manufactured with small clearances at an economic cost. Internal leakages have been reduced to a small fraction of their values in earlier designs. Screw compressors are therefore efficient, compact, simple and reliable. Consequently, they have largely replaced reciprocating machines in industrial applications and in refrigeration systems.

Recent advances in mathematical modelling and computer simulation may be used to form a powerful tool for the screw compressor process analysis and design optimisation. Such models have evolved greatly during the past ten years and, as they are better validated, their value as a design tool has increased. Their use has led to a steady evolution in screw rotor profiles and compressor shapes which should continue in future to lead to further improvements in machine performance. Evidence of this may be seen in the publications by *Sauls, 1994* and *Fujiwara and Osada, 1995*. In order to make such computer models more readily accessible to designers and engineers, as well as specialists, the authors have developed a suite of subroutines for the purpose of screw machine design, *Hanjalic and Stosic, 1997*.

There are several criteria for screw profile optimization which are valid irrespective of the machine type and duty. Thus, an efficient screw machine must admit the highest possible fluid flow rates for a given machine rotor size and speed. This implies that the fluid flow cross-sectional area must be as large as possible. In addition, the maximum delivery per unit size or weight of the machine must be accompanied by minimum power utilization for a compressor and maximum power output for an expander. This implies that the efficiency of the energy interchange between the fluid and the machine is a maximum. Accordingly unavoidable losses such as fluid leakage and energy losses must be kept to a minimum. However, increased leakage may be more than compensated by greater bulk fluid flow rates. However, specification of the required compressor delivery rate requires simultaneous optimisation of the rotor size and speed to minimise the compressor weight while maximising its efficiency. Finally, for oil-flooded compressors, the oil injection flow rate, inlet temperature and position needs to be optimised. It follows that a multivariable minimization procedure is needed for screw compressor design with the optimum function criterion comprising a weighted balance between compressor size and efficiency or specific power.

A box simplex method was used here to find the local minima, which were input to an expanding compressor database. This finally served to estimate a global minimum. The database may be used later in conjunction with other results to accelerate the minimization.

2. GEOMETRY OF SCREW COMPRESSOR ROTORS

Screw machine rotors have parallel axes and a uniform lead and they are a form of helical gears. The rotors make line contact and the meshing criterion in the transverse plane perpendicular to their axes is the same as that of spur gears. A procedure to get the required meshing condition as described in *Stosic, 1998*. More detailed information on the envelope method applied to gears can be found in *Litvin, 1994*.

To start the procedure of rotor profiling, the profile point coordinates in the transverse plane of one rotor, and their first derivatives, must be known. This profile can be specified on either the main or gate rotors or in sequence on both. Also the primary profile may also be defined as a rack as shown in Fig 2.

A helicoid surface and its derivatives for the given rotor profile can be found from the transverse plane rotor coordinates,. The envelope meshing condition for screw machine rotors gives the meshing condition either numerically, if the generating curves are given on the compressor rotors, or directly, if the curves are given on the rotor rack. This enables a variety of primary arc curves to be used and basically offers a general procedure. Moreover, numerical derivation of the primary arcs permits such an approach even when only the coordinates of the primary curves are known, without their derivatives.

The following are the elements of the rack-generated 'N' profile. The primary curves are specified on the rack: D-C is a circle with radius r_3 on the rack, C-B is a straight line, B-A is a parabola constrained by radius r_1, A-H-G are trochoids on the rack generated by the small circles of radii r_2 and r_4 from the main and gate rotors respectively, G-E is a straight line and E-F and E-D are circles on the rack. A full description of the rack generation procedure and rotor geometry is given in *Stosic* and *Hanjalic, 1997*. Three rotor radii, r_1-r_3 and the gate rotor addendum r_0 are used as variables for the rotor optimisation.

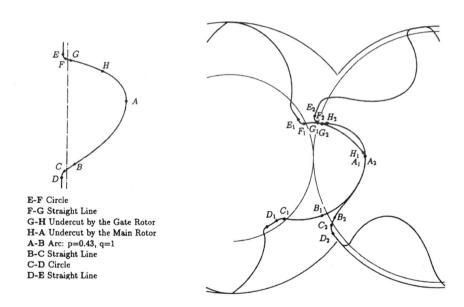

E-F Circle
F-G Straight Line
G-H Undercut by the Gate Rotor
H-A Undercut by the Main Rotor
A-B Arc: p=0.43, q=1
B-C Straight Line
C-D Circle
D-E Straight Line

Fig 2. Distribution of generating profile curves on the rack for 'N' rotors

Full rotor and compressor geometry, like the rotor throughput cross section, rotor displacement, sealing lines and leakage flow cross section, as well as suction and discharge port coordinates are calculated from the rotor transverse plane coordinates and rotor length and lead. They are later used as input parameters for calculation of the screw compressor thermodynamic process. For any variation of input parameters r_0 to r_3, the primary arcs must be recalculated and a full transformation performed to obtain the current rotor and compressor geometry. The compressor built-in volume ratio is also used as an optimisation variable.

3. COMPRESSOR THERMODYNAMICS IN OPTIMISATION CALCULATIONS

The algorithm of the thermodynamic and flow processes used is based on a mathematical model comprising a set of equations which describe the physics of all the processes within the screw compressor. The mathematical model describes an instantaneous operating volume, which changes with rotation angle or time, together with the equations of conservation of mass and energy flow through it, and a number of algebraic equations defining phenomena associated with the flow. These are applied to each process that the fluid is subjected to within the machine; namely, suction, compression and discharge. The set of differential equations thus derived cannot be solved analytically in closed form. In the past, various simplifications have been made to the equations in order to expedite their numerical solution. The present model is more comprehensive and it is possible to observe the consequences of neglecting some of the terms in the equations and to determine the validity of such assumptions. This provision gives more generality to the model and makes it suitable for other applications.

A feature of the model is the use energy equation in the form which results in internal energy rather than enthalpy as the derived variable. This was found to be computationally more convenient, especially when evaluating the properties of real fluids because their temperature and pressure calculation is not explicit. However, since the internal energy can be expressed as a function of the temperature and specific volume only, pressure can be calculated subsequently directly. All the remaining thermodynamic and fluid properties within the machine cycle are derived from the internal energy and the volume and the computation is carried out through several cycles until the solution converges. A full description of the model is given in *Hanjalic and Stosic, 1997*.

The following forms of the conservation equations have been employed in the model. The conservation of internal energy is:

$$\omega\left(\frac{dU}{d\theta}\right) = \dot{m}_{in}h_{in} - \dot{m}_{out}h_{out} + Q - \omega p \frac{dV}{d\theta}$$

where θ is angle of rotation of the main rotor, $h = h(\theta)$ is specific enthalpy, $\dot{m} = \dot{m}(\theta)$ is mass flow rate $p = p(\theta)$, fluid pressure in the working chamber control volume, $\dot{Q} = \dot{Q}(\theta)$, heat transfer between the fluid and the compressor surrounding, $\dot{V} = \dot{V}(\theta)$ local volume of the compressor working chamber. In the above equation the index in denotes inflow and the index out the fluid outflow.

The mass continuity equation is:

$$\omega \frac{dm}{d\theta} = \dot{m}_{in} - \dot{m}_{out}$$

The instantaneous density $\rho = \rho(\theta)$ is obtained from the instantaneous mass m trapped in the control volume and the size of the corresponding instantaneous volume V as $\rho = m/V$.

The suction and discharge port flow is defined by the velocity through them and their cross section area. The cross-section area A is obtained from the compressor geometry and it was considered as a periodical function of the angle of rotation θ.

Leakage in a screw machine forms a substantial part of the total flow rate and plays an important role because it affects the delivered mass flow rate and hence both the compressor volumetric and adiabatic efficiencies.

$$\dot{m}_l = \rho_l w_l A_g = \sqrt{\frac{p_2^2 - p_1^2}{a^2 \left(\zeta + 2\ln \frac{p_2}{p_1} \right)}}$$

Injection of oil or other liquids for lubrication, cooling or sealing purposes, modifies the thermodynamic process in a screw compressor substantially. Special effects, such as gas or its condensate mixing and dissolving in or coming out of the injected fluid should be accounted for separately if they are expected to affect the process. In addition to lubrication, the major purpose for injecting oil into a compressor is to cool the gas.

Flow of the injected oil, oil inlet temperature and injection position are additional optimisation variables if the oil-flooded compressors are in question.

The solution of the equation set in the form of internal energy U and mass m is performed numerically by means of the Runge-Kutta 4th order method, with appropriate initial and boundary conditions. As the initial conditions were arbitrary selected, the convergence of the solution is achieved after the difference between two consecutive compressor cycles is sufficiently small.

Numerical solution of the mathematical model of the physical process in the compressor provides a basis for a more exact computation of all desired integral characteristics with a satisfactory degree of accuracy. The most important of these properties are the compressor mass flow rate $\dot{m}$ [kg/s], the indicated power P_{ind} [kW], specific indicated power P_s [kJ/kg], volumetric efficiency η_v, adiabatic efficiency η_a, isothermal efficiency η_t and other efficiencies, and the power utilization coefficient, indicated efficiency η_i.

4. OPTIMIZATION OF THE ROTOR PROFILE AND COMPRESSOR DESIGN

The power and capacity of contemporary computers is only just sufficient to enable a full multivariable optimisation of both the rotor profile and the whole compressor design to be performed simultaneously in one pass. Nine optimization variables were used in the calculation presented, radii r_0, r_1, r_2, and r_3 were four rotor profile parameters, built–in volume ratio is another compressor geometry variable, compressor speed is an operating variable and oil flow, temperature and injection position are oil optimisation parameters.

A box constrained simplex method was used here to find the local minima. The box method stochastically selects a simplex, which is a matrix of independent variables and calculates the optimisation target. This is later compared with those of previous calculations and then their minimization is performed. One or more optimisation variables may be limited by the calculation results in the constrained Box method. This gives additional flexibility to the compressor optimisation.

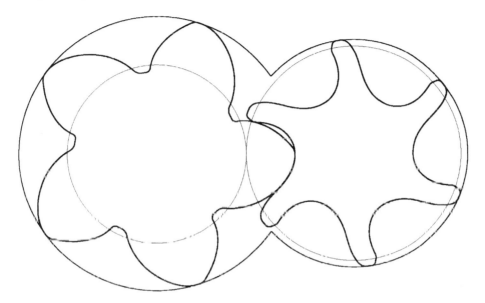

Fig. 3 Rotor profile optimized for an oil-free air compressor duty

The optimisation results, after being input to an expandable compressor database, finally served to estimate a global minimum. The database may be used later in conjunction with other results to accelerate the minimization.

The suction and discharge pressures were 1 – 3 bar for the dry air compressor and 1 - 8 bar for the oil flooded compressor, while the evaporation and condensation temperatures were 5 and 40 °C for R-134A. The centre distance and male rotor outer diameters were kept constant for all compressors, 90 and 128.450 mm respectively.

The optimisation criterion was the lowest compressor specific power. As a result, three distinctively different rotor profiles were calculated, one for oil-free compression and the other two for oil-flooded air and refrigeration compression. They are presented in Figs. 4-6.

Although the profiles somewhat look alike, there is a substantial difference between their geometry which is given in the following table as well as further results of the compressor optimisation.

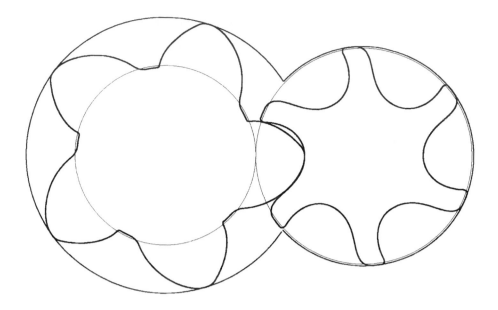

Fig. 4 Rotor profile optimized for an oil-flooded air compressor

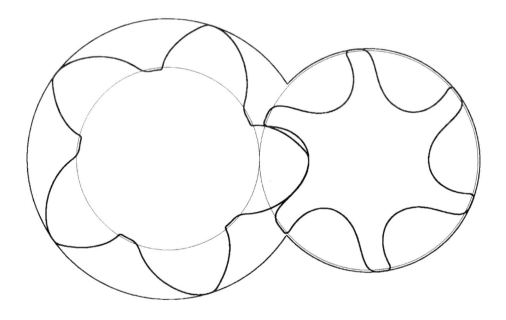

Fig. 5 Rotor profile optimized for a refrigeration compressor duty

   C603/001/2003

Table 1: Results of optimisation calculations for dry and oil flooded air compressors and oil flooded refrigeration compressor

	DryAir	Oil-Flooded Air	Refrigeration
r_0 [mm]	2.62	0.74	0.83
r_1 [mm]	19.9	17.8	19.3
r_2 [mm]	6.9	5.3	4.5
r_3 [mm]	11.2	5.5	5.2
Built-in volume ratio	1.83	4.1	3.7
Rotor speed [rpm]	7560	3690	3570
Oil flow [lit/min]	-	12	8
Injection position [°]		- 65	61
Oil temperature [°]		- 33	32

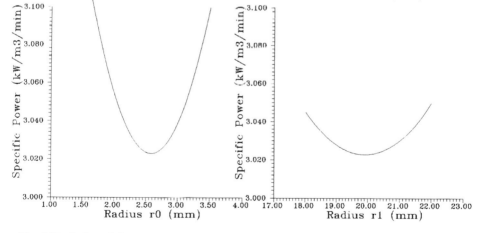

Fig. 6 Variation of the compressor specific power as function of the compressor rotor parameters

As an example, if the gate rotor addendum is analysed in detail, it can be concluded that, the size of the rotor blow-hole area is proportional to the addendum. Therefore r_0 should be made as small as possible in order to minimise the blow-hole. It would therefore appear that ideally, r_0 should be equal to zero or even be 'negative'. However, reduction in r_0 also leads to a decrease the fluid flow cross-sectional area and hence a reduction in the flow rate and the volumetric efficiency. It follows that there is a lower limit to the value of r_0 to obtain the best result. More details of single variable optimisation of screw compressor rotors can be found in *Hanjalic and Stosic, 1994.*

As in the case of any result of multivariable optimization, the calculated screw compressor profile and compressor design parameters must be considered with the extreme caution. This is because multivariable optimisation usually finds only local minima, which may not necessarily be globally the best optimisation result. Therefore, extensive calculations should be carried out before a final decision on the compressor design is made.

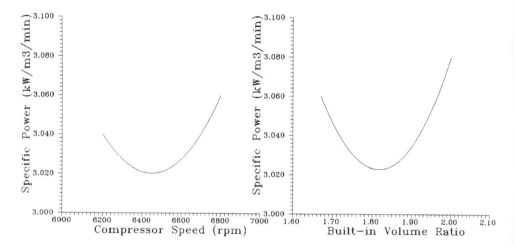

Fig. 7 Specific power as function of the compressor built-in volume and speed

The dry air compressor was chosen for further analysis. This is because the compression process within it is close to that of an ideal gas compressed adiabatically in which γ, the isentropic exponent, has the relatively large value of approximately 1.4. As an example of how the optimisation variables influence the compressor specific power, the radii r_0-r_3 are considered. The influence of the gate rotor tip addendum r_0, and the gate rotor radius r_3 are presented in Fig. 6, as well as the main rotor radii r_1 and r_2. In Fig. 7, the influence of the compressor built-in volume ratio, as well as compressor speed is presented.

5. CONCLUSIONS

A full multivariable optimisation of screw compressor geometry and operating conditions has been performed to establish the most efficient compressor design for any given duty. This has been achieved with a computer package, developed by the authors, which provides the general specification of the lobe segments in terms of several key parameters and which can generate various lobe shapes. Computation of the instantaneous cross-sectional area and working volume could thereby be calculated repetitively in terms of the rotation angle. A mathematical model of the thermodynamic and fluid flow process is contained in the package, as well as models of associated processes encountered in real machines, such as variable fluid leakages, oil flooding or other fluid injection, heat losses to the surrounding, friction losses and other effects. All these are expressed in differential form in terms of an increment of the rotation angle. Numerical solution of these equations enables the screw compressor flow, power and specific power and compressor efficiencies to be calculated.

A rack generated profile in 5/6 configuration rotors of 106 mm was used as an example to show how optimisation may permit both better delivery and higher efficiency for the same tip speed. Several rotor geometrical parameters, namely the main and gate tip radii, as well as the compressor built-in ratio and compressor speed and oil flow and temperature and injection position are used as optimisation variables and applied to the multivariable optimisation of the machine geometry and its working parameters for a defined optimisation target. In the case of the example given, this was minimum compressor specific power. It has thereby been shown that for each application, a different rotor design is required to achieve optimum performance.

REFERENCES

Fujiwara M, Osada Y, 1995: Performance Analysis of Oil Injected Screw Compressors and their Application, Int J Refrig Vol 18, 4

Hanjalic K, Stosic N, 1994: Application of mathematical modeling of screw engines to the optimization of lobe profiles, Proc. VDI Tagung "Schraubenmaschinen 94" Dortmund VDI Berichte Nr. 1135

Hanjalic K, Stosic N, 1997: Development and Optimization of Screw Machines with a Simulation Model, Part II: Thermodynamic Performance Simulation and Design, ASME Transactions, Journal of Fluids Engineering, Vol 119, p 664

Litvin F.L, 1994: Teoria zubchatiih zaceplenii (Theory of Gearing), 1956, Nauka Moscow, second edition, 1968, also Gear Geometry and Applied Theory Prentice-Hill, Englewood Cliffs, NJ 1994

Sauls J, 1994: The Influence of Leakage on the Performance of Refrigerant Screw Compressors, Proc. VDI Tagung "Schraubenmaschinen 94", Dortmund VDI Berichte 1135

Stosic N, Hanjalic K, 1997: Development and Optimization of Screw Machines with a Simulation Model, Part I: Profile Generation, ASME Transactions, Journal of Fluids Engineering, Vol 119, p659

Stosic N, 1998: On Gearing of Helical Screw Compressor Rotors, Proceedings of IMechEng, Journal of Mechanical Engineering Science, Vol 212, 587

C603/028/2003

CFD modelling and design optimization of a gerotor pump

F IUDICELLO
Hobourn Automotive, A Division of Dana Automotive Limited, Rochester, UK

ABSTRACT

Gerotor pumps are widely used in the automotive industry for fuel lift and injection, engine oil lubrication and in transmission systems. The CFD analysis of gerotor pumps can help optimise flow performance and reduce cavitation damage and noise. In this paper, a CFD design optimisation of a fuel lift gerotor pump has been conducted using a 3-D transient model with moving and deforming boundaries. The effect of different design parameters, such as porting geometry and rotor clearances has been investigated. Results showed that the rotor clearances and the inlet grooves are the most influential parameters for pump performance, while for noise and cavitation damage, porting sealing are most important. The optimum pump design was verified using experimental data

NOTATION

Cl_f	Face-to-face clearance [μm]	θ_1, theta1	Inlet major sealing angle [degree]
Cl_t	Tip-to-tip clearance [μm]	θ_2, theta2	Outlet major sealing angle [degree]
Co	Courant Number	θ_3, theta3	Outlet minor sealing angle [degree]
IG	Inlet groove	θ_4, theta4	Inlet minor sealing angle [degree]
OG	Outlet groove	ρ	Fluid density [Kg/m^3]
ν	Fluid kinematic viscosity [cSt]		

1 INTRODUCTION

Gerotor (GEnerated ROTOR) pumps are internal rotary positive displacement pumps in which the outer rotor has one tooth more than the inner, and utilises cycloidal gear profiles; the inner gear is the conjugate shape to the outer gear. Gerotor pumps can be used for pressures up to 200 bar. In automotive transmission applications, displacements can reach 41 cc/rev and pressures 21bar (1).

In order to optimise pump performance and reduce cavitation damage and noise, one must analyse the fluid dynamics inside the pump flow passageways. In a gerotor pump, the fluid is sucked into the inlet port through the inlet passageways and shifted from the inlet to the outlet port. Due to the rotor tip-to-tip and face-to-face clearances, flow leakage occurs between the high-pressure and low-pressure sides of the pump. To limit pressure, excess fluid is re-circulated to the inlet port through a pressure relief valve. The flow through the rotor clearances creates high fluid velocity and localised low-pressure areas, which produce air/vapour bubbles hence causing cavitation damage and noise.

CFD analysis can be used as a cost-effective design tool for the optimisation of pump flow performance and reduction of fluid borne noise (2,3). In order to optimise gerotor pump design, a realistic CFD model is required which takes into account gear meshing, leakage flow across clearances and cavitation bubble formation and collapse. 3-D transient CFD studies on gerotor pumps have been reported in the published literature, which use simplified modelling for flow leakage paths and deformation of pumping chambers (4-7). Jiang et al. (4), for example, did not consider the flow leakage paths; Haworth et al. (5) and Broberg (6) considered non-realistic clearances; and Lee et al. (7) considered realistic clearances and moving and sliding mesh, but did not model the time-dependent wall deformation. The Lee et al. model (7) can predict pump leakage and efficiency accurately; however, because it does not model local flow dynamics, it cannot predict cavitation damage or noise.

The CFD model developed in this paper can predict cavitation bubble formation, recompression and collapse, by realistically modelling the dynamics of gear rotation, meshing and sliding over the inlet and outlet ports, and flow leakage through the rotor set clearances. A preliminary validation of the CFD calculations has been carried out in (8). In this paper, the design optimisation of a diesel fuel lift pump for truck engines has been carried out. Similar analyses could be conducted for gerotor pumps with smaller or greater rotors, with a wide range of rotor clearances, pressure and speed operating conditions, and fluid type and viscosity. A Design Of Experiment (DOE) method was used to optimise pump performance, flow fluctuation and cavitation bubble formation. The CFD model and the optimisation procedure are described in Section 2. The optimisation results are discussed in Section 3 and summarised in Section 4.

2 METHODS

2.1 CFD Modelling

A 3-D view of the pump flow passageways and rotor set is shown in Fig. 1. Figure 2 shows the flow leakage paths through the rotor clearances. A CFD model of the pumping chambers and the inlet and outlet ports and pipes was generated (see Fig.3). A 3-D transient model was developed specially for gerotor pumps. The grid generation and time dependent manipulation was carried out using the pre-processor of STAR-CD ProStar (9). The mesh motion/rotation and deformation of the pumping chambers was defined using a "script". Arbitrary sliding interfaces were used to connect the rotating pumping volumes with the stationary inlet and outlet ports. The pressure relief valve was not modelled for this study.

The fluid properties of the calibration fluid at 40°C were used. This fluid is used for the end-of-line production tests to simulate diesel fuel at 70°C. The fluid was assumed to be incompressible and have density $\rho = 820$ Kg/m^3, and kinematic viscosity $v = 2.5$ cSt. The test conditions in Table 1 were considered. Constant pressure boundaries were used at the inlet and outlet pipes. The pump speed was varied from 170 to 3000 rpm (see Table 1). At these operating conditions the pressure relief valve is closed.

Table 1. Summary of the analysed test conditions.

Test Condition No.	Speed (rpm)	Outlet Pressure (bar)
1	170	0.5
2	1000	5
3	3000	5

2.2 Optimization Procedure

A Design Of Experiments (DOE) optimization procedure of the gerotor pump was carried out. The eight design parameters described in Table 2 and Fig. 4 were investigated. Three levels were selected for most variables to take into account non-linearity effects. The underlined levels in Table 2 indicate the nominal values of the factors. An L_{18} Matrix was used for the DOE Tests (10). Interactions between variables were assessed preliminarily. It was found that some interaction only existed between the relief groove OG (1.3 mm x 1.4 mm) and the outlet major sealing angle θ_2. To estimate this interaction separately, the two variables were placed in the first two columns of the DOE matrix. The interactions of the rest of the variables were equally distributed to all columns and, therefore, could not be estimated.

Table 2. Description of the design parameters and their levels.

Variable No.	Variable Description	Variable Levels
1	Relief groove OG	No Yes
2	Outlet major sealing angle θ_2 [degree]	22.5 - 33.75 - 45
3	Inlet major sealing angle θ_1 [degree]	22.5 - 33.75 - 45
4	Outlet minor sealing angle θ_3 [degree]	22.5 - 33.75 - 45
5	Inlet minor sealing angle θ_4 [degree]	22.5 - 33.75 - 45
6	Inlet Grooves IG	Yes – No
7	Face-to-face Clearance Cl_f [μm]	0 - 15 – 30
8	Tip-to-tip Clearance Cl_t [μm]	80 – 100 - 120

After conducting the DOE CFD tests, the data analysis was carried out to find the optimum set of parameters. The outputs described in Table 3 were considered. The analysis of means (10) was conducted to estimate the main effects of the 8 variables. These were defined, for each variable at a given level, as the average output at that level, minus the overall average output (10). The optimum value of each output was predicted using an additive model and verified experimentally. The flow fluctuation (flow ripple) was measured using the in-house Fluid-Borne Noise (FBN) Test Rig (11). This Test Rig uses a procedure based on the "secondary source method" developed by Bath University (12,13).

Table 3. Output description.

Output No.	Output Description
1	average outlet flow rate at test condition 1
2	average outlet flow rate at test condition 2
3	average outlet flow rate at test condition 3
4	RMS value of the outlet flow fluctuation at test condition 2
5	RMS value of the outlet flow fluctuation at test condition 3
6	minimum static pressure over 8 pumping cycles at condition 3

3 RESULTS AND DISCUSSION

The CFD results of the DOE analysis are in Figs 5 to 7. The main factor effects show that the most important variables for Outputs 1 to 3 are the tip-to-tip and the face-to-face clearances, and the inlet groove (see Figs 5, 6a, and 7a). These results seem to indicate that the major contributors to flow leakage are the rotor clearances and the presence of the inlet groove. The most important factors for the RMS value of flow fluctuation depend on speed. At low speed (1000 rpm, Output 4), the tip-to-tip clearance Cl_t, the inlet minor sealing angle θ_4, and the outlet major sealing angle θ_2 are the most important factors (Fig. 6b). At high speed (3000 rpm, Output 5), Cl_t, θ_2, the inlet major sealing angle θ_1, and θ_4 are the most important factors (Fig. 7b). The most important factors for Output 6 are θ_4, the rotor clearances, and θ_2 (Fig. 7c).

The optimum set of factors levels for the six outputs are in Table 4. Four optimum level combinations were selected for experimental verification to take into account manufacturing and design feasibility of some variables, see Table 5. As can be seen in Table 4, in general the optimum levels depend on the optimised output. However, the optimal levels for variables 1 and 3 are consistent for all outputs: a relief groove not present and $\theta_1=45°$. This seems to indicate that a relief groove of the geometry considered in this study (1.3 mm x 1.4 mm) not only contributes to the flow leakage and hence reduces flow performance, but also it does not reduce flow fluctuation or cavitation damage. Similarly, θ_1 should be kept at 45° to reduce leakage, flow ripple and cavitation damage. The optimum level for θ_2 is 3 ($\theta_2=45°$), at low speed conditions (Outputs 1, 2 and 4), and 1 ($\theta_2=22.5°$) at high speed condition (Outputs 3, 5 and 6). This seems to indicate that a large θ_2 helps reduce leakage, but at high speed is not optimal for noise and cavitation damage; however, at high speed the importance of θ_2 is low. The optimum level for θ_3 depends on the optimised output. As this variable is never important, it was decided to set its optimum level to 2 in the experimental verification.

Table 4. Optimum set of factors levels for the six outputs.

Variable No.	Variable Levels					
	Output 1	Output 2	Output 3	Output 4	Output 5	Output 6
1	1	1	1	1	1	1
2	3	3	1	3	1	1
3	3	3	3	3	3	3
4	3	3	1	1	2	3
5	1	1	1	3	3	1
6	3	3	3	3	3	2
7	1	1	1	2	1	3
8	1	1	1	1	1	3

Table 5. Selection of optimum level combinations.

Model No.	Variable No.							
	1	2	3	4	5	6	7	8
Opt-1	1	3	3	2	1	2	1	1
Opt-2	1	3	3	2	1	2	1	2
Opt-3	1	3	3	2	1	1	2	2
Opt-4	1	3	1	2	1	1	2	2

C603/028/2003 © IMechE 2003

The optimum level for θ_4 varies with the output considered. For Outputs 1 to 3, and 6, the optimum level is 1 (θ_4=22.5°), i.e. a larger inlet port helps the gear filling. However, the optimal level for Outputs 4 and 5 is 3 (θ_4=45°). As the noise issue is of secondary importance for this pump, level 1 was selected for this variable in the experimental verification. The optimum level for the face-to-face clearance Cl_f, is 1 (0.0 µm) for most outputs, and hence this was chosen for the models Opt-1 and Opt-2. However, because of the difficulty of having a tight tolerance in production, nominal values were considered in model Opt-3 and Opt-4. Similarly for Cl_t, the optimum level is 1 (80.0 µm) for most outputs. Nominal values were considered in models Opt-2 to Opt-4. In addition, the effect of having the nominal design inlet port with the remaining set of optimised levels was analysed in model Opt- 4.

A validation of the CFD results for the nominal model (see Table 6) was carried out in (8), where it was found that CFD calculations in general under-predict average outlet flow. A discrepancy was found between the CFD results and the experimental measurements of flow ripple. This discrepancy is still present for this study but the predicted trends seem to be valid for optimisation.

The results in Table 6 show that model Opt-1 can produce an improvement in flow performance of up to 44% (Output 1) over the nominal model. Because of the difficulty in achieving tight tip-to-tip clearance in production, model Opt-1 could not be verified experimentally. The CFD improvement for model Opt-1 reduces with speed (see Fig. 8a): 42% for condition 2 and 13% for condition 3. A reduction in the flow RMS (Outputs 4 and 5) is also predicted, especially at 1000 rpm. However, an increase in cavitation damage could be observed due to a greater amount of bubbles formed (lower minimum static pressure). 600 hours life tests are in progress to confirm these findings. A more accurate prediction of cavitation bubble formation could have been obtained if the cavitation model available in STAR-CD were used. Because of the time required for the calculation of this model, this was not practical for this study. Work is underway at STAR-CD software to reduce the convergence time for the cavitation model.

Table 6. Summary of the experimental verification.

Model No.	Output 1 (l/min)	Output 2 (l/min)	Output 3 (l/min)	Output 4 (l/min)	Output 5 (l/min)	Output 6 (bar)
Nominal CFD	0.193	1.070	6.340	0.116	0.179	-0.796
Nominal Exp.	0.250	1.367	6.440	0.037	0.062	(**)
Opt-1 CFD	0.361	1.949	7.259	0.063	0.145	-1.058
Opt-1 Exp.	(*)	(*)	(*)	(*)	(*)	(**)
Opt- 2 CFD	0.298	1.585	6.892	0.065	0.263	-0.764
Opt- 2 Exp.	0.298	1.66	6.53	0.050	0.109	(**)
Opt- 3 CFD	0.28	1.52	6.84	0.06	0.18	-0.670
Opt- 3 Exp.	0.267	1.51	6.53	(*)	(*)	(**)
Opt- 4 CFD	0.235	1.313	6.678	0.188	0.224	-0.888
Opt- 4 Exp.	0.279	1.367	6.683	0.062	0.112	(**)

(*) Data not available   (**) Not measurable

The predicted improvement in flow performance decreases from model Opt-1 to Opt-4. The flow reduces about 20% from model Opt-1 to Opt-2 for Outputs 1 and 2, and about 5% for Output 3. The flow reduction from model Opt-2 to Opt-3 is up to 6%, and up to 16% from model Opt-3 to Opt-4. Similar trends can be seen in the experimental measurements (see

Fig.8b) only for Output 2 (flow improvement of models Opt-2 to -4 gradually reducing). For Outputs 1 and 3, model Opt-1 performs best, and model Opt-3 performs worst. However, models Opt-2 to -4 still perform better than the nominal for all outputs. The lack of improvement in RMS flow values (Outputs 4 and 5) is also seen in the experiments. Further CFD development work will be required to improve the accuracy of flow ripple predictions.

4 CONCLUSIONS

A design optimization of a gerotor pump for vehicle fuel lift applications has been conducted. A DOE method has been used to optimise the levels of eight design parameters. The DOE analysis of the CFD results has shown that the most important parameters for the flow performance are the rotor clearances and the inlet grooves. The most important variables for the flow fluctuations RMS are the tip-to-tip clearance, the inlet minor sealing angle and the outlet major sealing angle. Based on the importance of the optimised outputs, four optimum parameter sets were selected for experimental verification. It was verified that the optimum models could improve flow performance. The increase in RMS flow fluctuations predicted by the CFD calculations was also verified experimentally. Further work is required to improve the CFD predictions of flow fluctuation and to include cavitation modelling.

REFERENCES

(1)   Gerotor selection and pump design. Nichols Portland, A division of Parker Hannifin Corporation.
(2)   Iudicello, F. and Baseley, S. CFD modelling of the flow control valve in a hydraulic pump. *PTMC 1999*, Bath University, Burrows, C.R. & Edge, K.A. eds, 1999, 297-312.
(3)   Wang, L., Fernholz, C.M. and Bishop, L. Hydraulic steering pump cavity flow CFD simulation to improve NVH performance. Proceedings of the 2001 Noise and Vibration Conference, SAE 2001-01-1611, 2001, Transverse City, Michigan.
(4)   Jiang, Y., Przekwas, J., Perng, C-Y. Computational analysis of oil pumps with an implicit pressure based method using unstructured mixed element grid. SAE 960423.
(5)   Haworth, D.C. Maguire, J.M., Rhein R. and ElTahry, S.H. Dynamic fluid flow analysis of oil pumps. SAE 960422, 1996.
(6)   Broberg, R. New moving grid feature enables CFD simulation of internal gear pump. Internal Report, CFX/AEA Technology Engineering Software, 2001, Canada.
(7)   Lee W.-T., Yang, B.G., Dong, M., Yu, D. and Chen G. Computational Fluid Dynamics of automobile gerotor oil pump. SAE 970681, 1997.
(8)   Iudicello, F. and Mitchell D. CFD modelling of the flow in a gerotor pump. *PTMC 2002*, Bath University, Sept 2002.
(9)   STAR-CD User Guide Version 3.15, 2001, Computational Dynamics Ltd.
(10)   Phadke, M.S. Quality engineering using robust design. Prentice Hall, 1989.
(11)   Iudicello, F. and Baseley, S. Fluid borne noise characteristics of hydraulic and electro-hydraulic pumps. *PTMC'99*, 1999, pp. 313-323.
(12)   Edge, K.A. and Johnston, D.N. The 'secondary source' method for the measurement of pump pressure ripple characteristics. Part 1: description of method. *Proc. Instn Mech Engrs*, 1990, (204): 33-40.
(13)   ISO Standards Hydraulic fluid power - Determination of pressure ripple levels generated in systems and components-Part1: Precision method for pumps. ISO 10767-1, 1996.

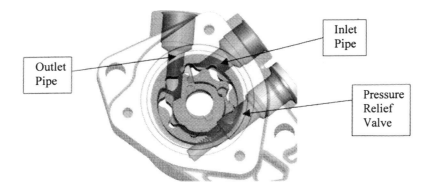

Figure 1. 3-D view of the pump flow passageways and rotor set.

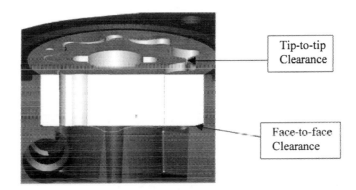

Figure 2. Detail of the flow leakage paths.

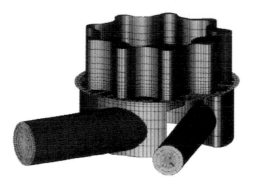

Figure 3. CFD model of the pumping chambers, ports, and inlet and outlet pipes.

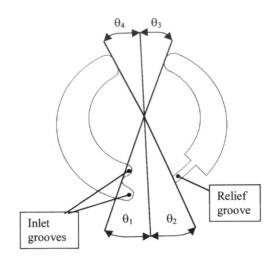

Figure 4. Design parameters for the porting geometry.

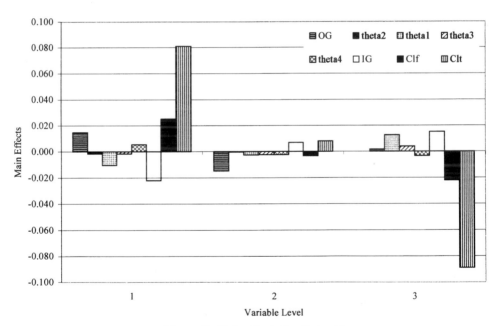

Figure 5. Main effects for Output 1.

C603/028/2003 © IMechE 2003

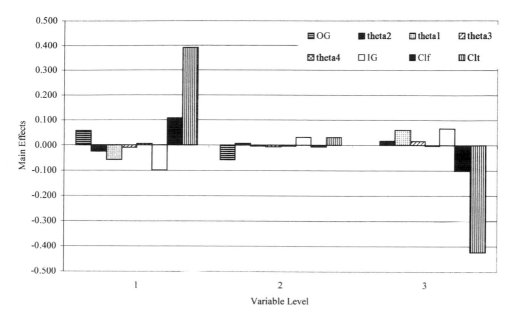

(a)

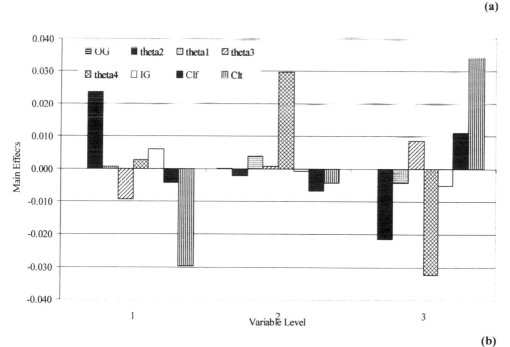

(b)

Figure 6. Main effects for (a) Output 2, (b) Output 4.

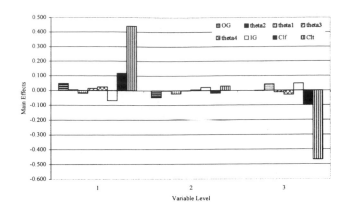

(a)

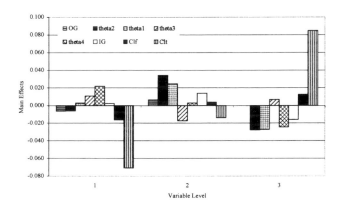

(b)

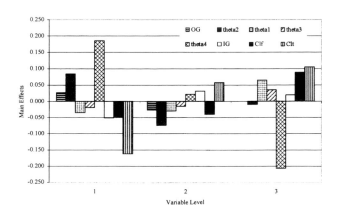

(c)

Figure 7. Main effects for (a) Output 3, (b) Output 5, (c) Output 6.

C603/028/2003 © IMechE 2003

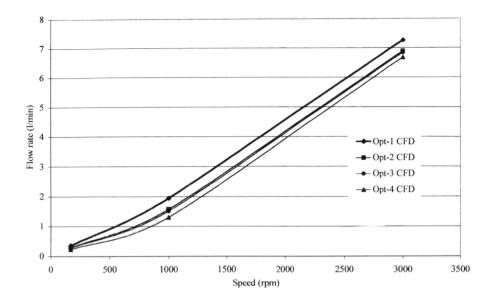

(a)

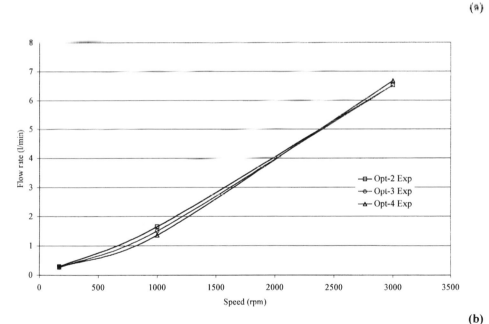

(b)

Figure 8. Output flow rate vs speed: (a) CFD results, (b) experimental measurements.

C603/015/2003

Industrial centrifugal compressors – physical background and practice of modern gas dynamic design

Y GALERKIN, V MITROFANOV, K DANILOV, and E POPOVA
Compressor Department, TU Saint-Petersburg, Russia

SYNOPSIS

Industrial centrifugal compressor gas dynamics are the focus of our Compressor Department activity. D ecades of experiments gave vast information on the correlation of performances with geometry and gas dynamic factors. The combination of measurements inside rotating impellers and flow visualization clarified important phenomena that influence head input and head losses. The developed physical model takes into account negative and positive secondary flow influence, wake formation in the corner "blade suction side – shroud" while flow rate is equal or less than design, separation absence on a shroud, etc. Surface velocities' diagrams (non-viscid flow) adequately describe the blade shape's strong influence on mechanical head input, optimal flow rate and head losses. In the design method for a stage, all calculations are non-dimensional.

The Mathematical model for performance prediction is based on the physical one and on empirical coefficients obtained from the vast number of experiments. Numerical optimization sufficiently diminishes the necessary number of costly model tests. The design of industrial compressors in the Compressor Department started in late seventies and is concentrated now on pipeline compressors. About 250 units of two dozen types (total power 3 400 000 kWt) operate in gas industry of 5 countries now. The new generation of Russian pipeline compressors based on the Compressor Department gas dynamic design, covers the range of power 3.0 – 25.0 Megawatt, pressure ratio 1.37 – 3.0, exit pressure 28 – 150 bar, number of stages 2 – 8, etc. Due to rational s chemes a nd flow path optimization their efficiency and range are superior in comparison with analogues.

NOMENCLATURE:

c - absolute velocity of flow (in m/s),

D_2 – impeller diameter (in m),

k - isentropic exponent,

$$M_U = \frac{u_2}{\sqrt{k * R * T}} \text{ - Mach number,}$$

$$M_w = \frac{w}{\sqrt{k * R * T}} \text{ - Mach number,}$$

$$Re_U = \frac{u_2 D_2}{\nu} = \frac{u_2 D_2 \rho_i}{\mu} \text{ - Reynolds number,}$$

Ro - Rossby number,

u - transition velocity (velocity of impeller rotation) (in m/s),

V_{inl} – inlet volumetric flow (in m^3/s) related to total parameters,

w - relative flow velocity (in m/s),

$\Phi = 0.785\ V_{inl} / (D_2^2\ u_2)$ - flow rate coefficient,

η - polytropic efficiency,

$\psi_T = c_{U2} / u_2$ - Euler head coefficient,

Subscripts: 2 – related to impeller diameter, u – tangential component of velocity.

1. INTRODUCTION

Centrifugal compressors play a very important role in modern industry. Their energy consumption is high. For instance, about 4200 powerful pipeline compressors of "Gazprom" Russia (about 45 000 000 kWt in total) use fuel with the cost of some billion dollars annually.

A century long development of industrial centrifugal compressors gave brilliant samples of perfect designs, produced sophisticated design methods, gave good understanding of flow phenomena. Sets of well-proven model stages give the possibility to design new compressors very quickly and with a high level of reliability. Though, there are at least two reasons why the process of new compressor (not based on existing model stages) and new model stages design continue:

- new demands of compressor users can not always be met by existing model stage sets,
- new ideas and methods give the possibility to create more effective model stages and compressor flow paths.

The goals of gas dynamic design are:

- to guarantee a given pressure ratio at a given flow rate,
- to obtain the maximum possible efficiency or (and) some other gas dynamic performance properties satisfying user's demands,

Both goals must be achieved inside mechanical, production and other constrains.

C603/015/2003 © IMechE 2003

Flow behavior in centrifugal compressors is quite complicated. The main reasons for this are:

- very high blade loads and low aspect ratio of blades create intensive secondary flows,

- strong deflection of flow by impeller blades creates intensive inertia force in the direction normal to flow. The normal inertia force leads to flow stabilization (suppression of turbulence fluctuations) on blade suction side. The same force leads to flow destabilization (intensification of turbulence fluctuations) on blade pressure side,

- sufficient flow deceleration in impellers and diffusers leads to flow separation at design and off-design flow rates. Low energy zones appear on impeller blades' suction side due to flow stabilization,

- numerous U-turns of flow paths distort flow field.

The usually applied methods of primary design are no more than some sets of more or less proven rules on how to find a proper flow path configuration. This approach can not guarantee satisfactory results as soon as these rules have an approximate and qualitative character. Therefore, experiments are necessary to prove design results. Some corrections could be made after experimental checking: to rise or lower pressure ratio at given flow rate, to eliminate possible mismatching of stage elements or stages in a multistage compressor, to make some improvements of a flow path. These experiments are still costly, in spite of certain progress in design methods [1].

The modern viscid codes give very useful information but they do not always lead to results that correlate with experimental data inside acceptable accuracy – [2], [3] The codes' application is too time consummating for any wide investigation and comparison of designed compressor variants.

The need for a reliable and not too complicated modeling instrument, was met by formulating the special approach that combines detailed physical model and statistical reduction of numerous experimental data. The so-called Universal Modeling method was created following intensive investigations executed by the TU SPb Compressor Department – [4], [5], [6], [7], [10].

2. METHODS OF INVESTIGATION

The Compressor Department started its activity on centrifugal compressors at the end of 1950'. Model tests for overall and elements' performance evaluation are combined with detailed study of flow structure within rotating impellers. Different methods of flow visualization are applied too. There are four electrically driven test rigs with maximal power 600 kWt and 18000 RPM (variable). One of the test rigs has the close loop with maximum pressure 100 bar. There are some test rigs of lesser power for specific applications - unsteady flow phenomena investigations, for instance. Detailed description of instrumentation and problems of flow measurement inside impellers, visualization and model tests' procedure in general could be found in [8], [9].

3. EXPERIMENTAL DATA ON FLOW BEHAVIOR [9][1]

Jet – wake flow. Direct measurements of total and static pressures inside impellers demonstrated existence of the low energy zones on the blade suction side at the exit of the impellers ("wake") at Φ_{des} and $\Phi < \Phi_{des}$. The typical measured relative velocity profiles on different radiuses (blade to blade) are presented at Fig.1. Measurements were executed in the impeller with the outer diameter $D_2 = 0.610$ m at the low – speed test rig ECC-2. The 2-D impeller with arc blades evidently did not belong to the most effective. The wake that was registered just after the end of blades (radius = 312 mm) leaves to the active part of flow ("jet") less than 2/3 of exit area of the channel. Evidently, it leads to significant mixing losses.

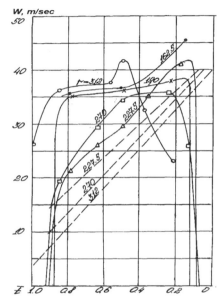

Fig. 1. Velocity profiles on different radiuses of the mean blade to blade surface (total and static pressures are measured by means of the pressure transducer). Low effective 2D impeller

The reason of wake formation is the negative effect of rotation on shear stresses. Normal inertia force (Coriolis force is its main component) suppresses flow turbulence pulsation in the direction normal to main flow at blades' suction side. Boundary layers became thicker and less resistant to opposite pressure gradients. The measured velocity profiles at the suction side of the blade at different radiuses are presented at Fig.2. The profiles are compared with two variants of boundary layer theory prediction – rotation neglected and taken into account.

---

[1] The results presented in this part belong to the Authors and to some other researchers who worked on the problems in the Compressor Department. The references on publications can be found in the book [9].

C603/015/2003 © IMechE 2003

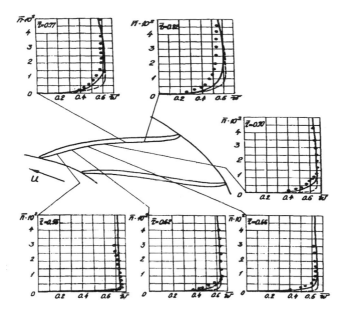

Fig. 1. Velocity profiles in the boundary layer on different radiuses of the mean blade to blade surface (total and static pressures are measured by means of the pressure transducer). High effective 2D impeller

The wake dimension and mixing losses can be diminished by proper design of blades. For instance, the impeller 085/065 was designed with constant velocity along the suction side except for the exit area where blade load disappears. The calculated non viscid and measured surface velocities are compared at Fig.3.

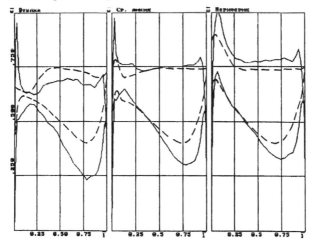

Fig.3 Measured (stroke lines) and calculated (solid lines) surface velocities at the blades of the high-effective 2-D impeller. Hub, mean and shroud blade to blade surfaces. Design flow rate

These velocities are calculated on a base of measured pressure distribution on blades and a total pressure of non-viscid flow in a relative motion. Non-viscid quasi-3D calculations correlate well with experimental data in the bigger part of blades. It shows that the wake formation starts at the very end of blades in this case. The impeller 085/065 demonstrated high efficiency.

The results of direct measurements are supported by flow visualization. Zones of a low energy were visualized by means of a thin powder. The powder b eing p ut i nto t he f low p ath w as noted to stick to surfaces where shear stresses are low. Typical results are presented at Fig.4.

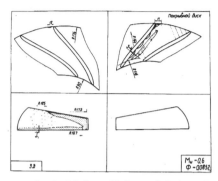

Fig. 4 Low energy zones in the high – efficiency 2-D impeller. Design flow rate.

When $\Phi=\Phi_{des}$ the wake near the hub was found only after $D/D_2>0.95$. But near the shroud it begins at $D/D_2=0.71$ i.e. the wake is of 3D character. It is the result of higher level of velocities and stronger deceleration near the shroud. To avoid noticeable flow separation is possible only in impellers with very low Euler head coefficient $\psi_T = 0.4 - 0.5$.

Pressure side of blades. Measurements and visualization demonstrated that flow separation never occurs there – at any incidence angles. When the flow rate is much greater than design, the peak of velocity deceleration leads to small separation bubble near the leading edge. This separation can not propagate due to the influence of normal inertia force. This effect is opposite to the one to described above in connection with the wake formation.

Split blades. The popular solution for 2-D and 3-D impellers is to apply two rows of blades – long and shortened ones. Reduction of blades at the inlet increases throat area that is good from any point of view. Unfortunately the usually applied short blades influence unfavorably the velocity diagram of the long ones. The velocity diagram – Fig.4 - demonstrates an inlet with significant positive incidence. The attempt to improve the situation by diminishing the inlet angle of short blades gave unexpected result. The performance comparison of a stage with two split impeller blades shows that the correction of the inlet angle lead to rise of efficiency but cut the range left and right from the design flow r ate. T he A uthors h ave n o positive solution for split blades at the moment.

C603/015/2003 © IMechE 2003

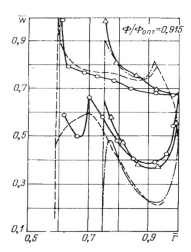

Fig.5 Calculated (stroke lines) and measured (solid lines) surface velocities on blades of some split impeller. Design flow rate.

Flow on shrouds and hubs. Visualization never demonstrated any separation at shrouds and hubs. Separation zones absence at shrouds, in spite of a sharp flow turn at an impeller inlet, is a result of shroud rotation. Special experiments with rotating channels (E.Smirnov, S.Iourkin, Gas Dynamic Dept. TU St.Petersburg) demonstrated the strongest influence of the Coriolis force on boundary layer particles. Due to secondary flow velocity component $w_{(sec)}$ the Coriolis force component appears with the same direction as a main flow. This component is proportional to $2*(u/r)*w_{(sec)}$.

Vaned diffuser. The results of powder visualization for vaned diffuser at $\Phi=\Phi_{des}$ demonstrated a separation zone existence on the pressure sides of blades (these sides are pressure sides in the pitch direction only). The Rossby numbers are positive at pressure sides of blades - inviscid core velocities are decreasing along the normal to them.

Similarity criteria. It is important to take into account the influence of similarity criteria k, M_U, Re_U and surface roughness. Regular experiments in the close-circuit high-pressure test rig are going on at the Compressor Department. Authors of the investigations have found that mechanically treated surfaces behave as hydraulically smooth up to $Re_U=8*10^9$ [9]. An important effect of Re_U is connected with sufficient decrease of disc friction losses for small specific speed stages.

The similarity criteria k influences flow through a density ratio in flow path cross sections. It is taken into account in a course of flow parameter calculations.

The experimental data demonstrates that between M_U, M_{W1} and M_{WMAX} values only the last one gives good correlation for η_{opt} change versus RPM for different impellers.

4. COMPUTER AIDED GAS DYNAMIC DESIGN [5], [6], [7], [9], [10]

CAD system is the set of computer codes for reverse (optimal design) and direct (performance prediction) applications. The gas dynamic mathematical model is based on the described understanding of flow behavior. Sets of algebraic equations take into account all main details of flow paths configurations, flow rates and similarity criteria values. The equations include empirical coefficients that are statistically proved by vast experimental data. Simplified flow description by algebraic equations reduces the time that is necessary for one stage efficiency calculation to some negligible value. It is possible to optimize main dimensions of a stage by comparison of numerous variants therefore. Blade passage configuration, leading edge shape and position, etc., are subjects of the analysis and qualitative optimization by quasi-3D calculations. The steps of compressor design are:

- instant analysis and optimization: number of shafts, stages, intercoolers, RPM of shafts or flow coefficients of stages, impeller head coefficients and peripheral speeds, main body dimensions. The variant's calculation is based on stages' efficiency approximate definition. The set of the proper equations is the product of statistic data on efficiency of stages [11]. The compressor stages' normalized (non dimensional) parameters are defined - Φ_{des}, $\psi_{T\,des}$, M_U, Re_U,

- optimization of stages main dimension by 500 - 800 (or more) variants' comparison,

- blade qualitative optimization on the basis of quasi-3D calculations,

- performance map calculation.

5. SAMPLE OF CAD SYSTEM APPLICATION

Main steps of design can be demonstrated on some arbitrary sample. Let us choose the object of design: the pipeline compressor 16 MWt, with pressure ratio $\Pi = 3$ for the installation at a well with reduced pressure 10 bar. The RPM is given by the driver choice – 5100 RPM.

1. Instant analysis of the general scheme. The proper code gives the instant answer on the expected main data versus RPM, number of stages, Euler head coefficient, etc. The comparison of variants with $\psi_{T\,des} = 0.6$ and different number of stages is presented in the table below as a sample. Variants with different $\psi_{T\,des}$ and reduced impeller diameters on last stages can be analyzed as well.

EXPECTED MAIN PARAMETERS OF THE PIPE LINE COMPRESSOR VERSUS
NUMBER OF STAGES

Num. Of stg.	η_{com}	η_{1st}	η_{lst}	D_2, m	Φ_1	Φ_{lst}	Body volume, (apprx.) m^3	u_2, m/c
5	0.803	0.825	0.774	1.017	0.046	0.023	3.35	271
6	0.820	0.829	0.803	0.916	0.060	0.030	3.47	245
7	0.827	0.829	0.820	0.764	0.077	0.038	3.56	225
8	0.825	0.807	0.828	0.790	0.094	0.047	3.68	211
9	0.827	0.827	0.30	0.751	0.1086	0.053	3.83	200

C603/015/2003 © IMechE 2003

It is up to a designer to choose the variant in accordance with all set of considerations – gas dynamic, structural mechanics, manufacturing. The variant with 6 stages is chosen for further calculations in our case.

2. Stage optimization. Comparison of hundreds variants gave the optimal dimensions to the each stage – inlet diameter, blade height at the inlet and exit of impellers, number of blades and their angle, etc. The meridian cross section of the flow path, velocity triangles, flow parameters in all control planes are ptresented at a computer monitor in a course of design.

3. Impeller blade optimization (qualitative). A specially developed code is a combination quasi – orthogonal and singularity methods. It treats flow field as non viscid and quasi three – dimensional.

4. Performance map calculation. The main dimensions of the flow path were analyzed and corrected, meaning manufacturing considerations, etc. The performance map for some range of RPM was calculated by the direct solution code CCPM-G4E-32. The results are presented in Fig.6.

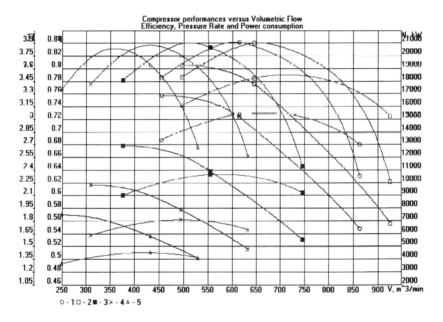

Fig. 6. **Performance map of the six stage pipeline compressor. RPM = 0.7, 0.8, 0.9, 1.0, 1.05 of the design ones 5100 1/min (code CCPM-G4E-32)**

6. DESIGN PRACTICE – UNIVERSAL MODELING APPLICATION

The compressors with flow paths designed by the Authors and their colleagues operate in refrigeration, petrochemical and gas industries. The pipeline and gas storage compressors are main objects of design. Two and a half hundred of gas industry compressors with 6.3 - 16 MWt of power and delivery pressure 28 – 125 bar are in constant production since 1980-s and operate successfully. The new generation of compressors was created in 1997 – 2002 and the process is gaining momentum. New Russian pipe – line compressors based on the Compressor Department gas dynamic design cover range of power 3.0 – 25.0 Megawatt, pressure ratio 1.37 – 3.0, exit pressure 56 – 150 bar, number of stages 2 – 8, etc. In accordance with industry specialists opinion these machines are superior to analogs in range and efficiency [12]. It is important to note that gas dynamic designs were realised with minimum model tests (or with none in some case). This demonstrates the validity of the applied design procedures.

REFERENCE

1. D. Japikse. Design system development for turbomachinery (turbopump) designs - 1998 and a decade beyond. JANNAF Conference. Cleveland, Ohio. July 15–17, 1998.

2. Gallus H.E. Recent Research Work on Turbomachinery Flow. Yokohama International Gas Turbine Congress, Yokohama, 1995.

3. Y.Galerkin, V.Mitrofanov, M.Geller, A.Toews. Experimental and numerical investigation of flow in an industrial centrifugal impeller, IMechE Conference transactions "Compressors and their systems", London, 2001.

4. Seleznev K.P., Galerkin I.B., 1982, "Centrifugal Compressors". Maschinostoenie, 1982. Leningrad, 271 p.(In Russian).

5. I.Galerkin, E.Popova. Industrial Centrifugal Compressors - Gas Dynamic Design and Optimisation Concepts, Modern Numerical Possibilities. VDI Berichte, N 1109, Aachen, 1994.

6. I.Galerkin, K.Danilov, E.Popova. Universal Modelling for Centrifugal Compressor Optimal Design. VDI Berichte, N 1208, Hannover, 1995.

7. Y.Galerkin, K.Danilov, E.Popova. Universal Modeling Method for Centrifugal Compressors – Gas Dynamic Design and Optimization Concepts and Application. Proceedings of the 1995 Yokohama International Gas Turbine Congress, 1995.

8. I.Galerkin, F.Rekstin. Methods of investigation of centrifugal compressors. "Mashinostroenie", Leningrad, 1969, 303 p. (In Russian).

9. Y.Galerkin (head editor). Transactions of the TU SPb in compressors. Saint-Petersburg, 2000, 443 p. (In Russian).

10. Y.Galerkin, K.Danilov, E.Popova. Design philosophy for industrial centrifugal compressors, IMechE Conference transactions "Compressors and their systems", London, 1999.

11. Popova E.I. Turbomachine stage optimization on a base of Math Modelling. Doctor's thesis. Saint-Petersburg Techn. University. 1991. (In Russian).

12. Vasiliev Y.S. (TU SPb), Rodionov P.I. ("GAZPROM"), Sokolovsky M.I. ("Iskra"). High-efficiency centrifugal compressors of new generation. Scientific basis, optimized design, manufacturing. "Industry of Russia", #10 – 11 (42 – 43), 2000, Moscow, p. 78 – 85. (In Russian)

C603/015/2003 © IMechE 2003

Turbo-Process Machinery

C603/017/2003

Exducer turbines, the optimized solution for liquefied gas expanders

H E HYLTON and **E H KIMMEL**
Ebara International Corporation, Cryodynamics Division, Sparks, Nevada, USA

1 ABSTRACT

The liquefaction process for natural gas requires LNG expanders with flexible operation regarding mass flow and differential pressure. The first pilot expanders were installed in Malaysia and were designed as hydraulic turbines with variable inlet guide vanes and air-cooled generators. Sealing problems, low reliability and hydraulic efficiency losses due to guide vane clearances resulted in the development of a completely different generation.

The second generation of LNG expanders, designed as a variable speed cryogenic turbine generator, eliminates the variable guide vane losses, but requires step down and step-up transformers and electronic frequency converters. For power generation in the range above 1.5 MW electronic frequency converters are less economical and constant speed cryogenic submerged turbine generators are the preferred solution. LNG expanders require a design that permits the turbine generators to adjust their operation according to a variety of input parameters.

The new and advanced LNG expanders use variable pitch Exducer Turbines operating at constant speed thus eliminating expensive transformers and frequency converters and increasing the total power output. Design and performance of the Exducer Turbine, the third generation of LNG expanders, is presented and compared with conventional variable speed cryogenic turbines. Exducer Turbines are the optimised solution for all applications in new liquefaction plants of any size and process, in floating LNG production units, for de-bottlenecking of existing LNG plants, and for two-phase expansion in specialized processes with partial LNG vaporisation.

2 INTRODUCTION

The liquefaction technology of natural gas involves a complex process with numerous systems interacting to produce the desired output. The conventional liquefaction process for natural gas and other hydrocarbons is to operate at a high pressure through the condensation phase, after which the high pressure of the condensed liquefied gas is reduced by expansion across a Joule-Thomson valve (1, 4).

The Joule-Thomson expansion is essentially a constant enthalpy process, in which the inlet and exit velocities of the fluid are equal and the heat transfer is negligible. A temperature rise of approximately 1-2 °C can be expected in the process liquid stream during this type of expansion.

Any temperature rise during the liquefaction process has an undesirable result and efforts are being taken in recent liquefaction plants to change the isenthalpic expansion into an almost isentropic expansion process. An isentropic expansion will result in a much lower exit temperature when compared to the Joule-Thomson expansion. Cryogenic LNG expanders are designed to achieve an LNG expansion process as close as possible to an isentropic thermodynamic process.

The first generation of LNG expanders with external generator performed with 75% isentropic efficiency (2). The second generation, the variable speed LNG expanders with submerged generator are operating at Oman LNG (7) with an isentropic efficiency of 82%. Further improvements in hydraulics and generator design resulted in isentropic efficiencies on the test stand of up to 88%. Cryogenic liquid expanders with high isentropic efficiencies cool down the LNG production stream by approximately 2-3 °C and increase proportionally to the enthalpy reduction the total LNG output and the overall plant profitability (4) (6). The payback time of LNG expanders is one of the shortest amongst all investments in petrochemical equipment.

3 EXTERNAL GENERATOR LNG EXPANDERS

The first generation of LNG expanders was installed in the mid-nineties at the National Helium Corporation in Liberal, Kansas, USA and at the MLNG-Dua plant in Bintulu, Sarawak, Malaysia. The design of these first experimental LNG expanders included a constant speed liquid turbine mechanically coupled to an external air-cooled generator. The turbines were fitted with semi-axial inflow runners and complex, mechanically adjustable inlet guide vanes to accommodate variations in performance conditions. The adjustable inlet guide vanes direct the fluid stream onto the runner and approximate the different fluid velocity vector angles brought about by changes in flow rate.

While these concepts were entrenched in water-power turbine designs, they were not innovative and did not avail themselves of established cryogenic rotating machine designs and the latest in variable speed hydraulic technology. The resultant machines were heavy, large and mechanically complex.

Figure 1 shows the first generation LNG expander at the MLNG-Dua plant in Bintulu, Malaysia. The extremely large dimensions of the expander require a platform of three floors. The design with external generator includes a heated double cryogenic mechanical shaft seal, mechanical shaft couplings between turbine and generator and a heavy duty thrust bearing. The isentropic efficiency of the first generation expanders was acceptable and reported with 75 %, but the downtime of the expanders were exceptionally high due to failures caused by the rudimentary design (2).

4 SUBMERGED GENERATOR LNG EXPANDERS

The design of the second generation of LNG expanders uses fixed geometry guide vanes, radial inflow turbine hydraulics and submerged induction generators operating at variable speed. The entire unit, both turbine and generator are completely immersed in the LNG fluid stream. This approach completely eliminates the need for any dynamic seals or couplings, and also eliminates heavy structure and the necessary alignment between turbine and generator.

Figure 2 compares size and weight between first and second-generation LNG expanders. The total weight is reduced to less than one third and the height is reduced to less than one half. Second generation LNG expanders are operating at Oman LNG in Sur, Oman and are currently being installed at MLNG-Tiga in Bintulu, Malaysia.

Figure 3 shows the installation of the variable speed LNG expanders at the construction site of MLNG-Tiga, Malaysia. The small dimensions of the expander require only a platform of one floor.

Figure 4 presents the basic design of the second generation LNG expanders showing three turbine stages, induction generator, rotating shaft, housing, power cables and containment vessel with inlet and outlet piping. The ball bearings are submerged and lubricated by the cryogenic fluid and a special thrust-balancing device is mounted between turbine and generator to increase the operational life of the bearings.

Submerged generator LNG expanders with fixed geometry guide vanes are operating at variable speed to adjust the expander performance to the requirements of the liquefaction process. Figure 7 shows the typical performance for variable speed LNG expanders as a function of differential head and flow. In case of no-load operation (8) with a de-energized generator the expander is operating along the no-load characteristic curve N. With increasing differential head or increasing flow the rotational speed increases. Should the rotor of the expander be prevented from rotating by some kind of device, the expander operates like an orifice without producing any power. The differential head across the expander follows a typical parabolic orifice curve L.

If the generator is energized and produces electrical power at a certain rotational speed and frequency, then the expander operates along a certain performance curve T_0 to the right side of N and above L. If the rotational speed of the expander increases or decreases, then the performance curves shift to higher (T_1) or lower (T_2) differential heads. All of the performance curves T_N for different rotational speeds N are approximately parallel to the orifice curve L and are shifted in the vertical direction of the differential head. This principle follows from the application of the affinity laws for hydraulic turbines and pumps. For any given values of flow and differential head within the range between the no-load and orifice characteristics N and L there is a certain rotational speed, which intersects with these given values. This feature enables the fixed geometry variable speed LNG expander to operate within a wide range of flows and differential heads.

The generator torque controls the variable speed of the LNG expander and the generator power frequency depends on the rotational speed. The variable frequency is then converted by an electronic frequency converter VSCF to the power net frequency of 50 Hz or 60 Hz. VSCF means Variable Speed Constant Frequency. For many years VSCF converters have been used for variable speed wind turbines, in steel rolling mills and for other industrial applications requiring full or partial regenerative capabilities. Cryogenic induction generators are more efficient at higher voltage. VSCF converters operate at medium voltage and it is necessary to connect the VSCF through step-down and step-up transformers to the generator and to the power net.

With increasing power the sizes and capital investments for VSCF converters are proportionally increasing with power, whereas for LNG expanders they are only increasing by the square root of the power. Therefore, variable speed LNG expanders operating with electronic frequency converters in the range above 1.5 MW are less economical and submerged generator LNG expanders operating at constant speed are the preferred solution.

5 EXDUCER TURBINES

The third generation of LNG expanders operates at a constant speed and eliminates the need for transformers and frequency converters. As stated above, the liquefaction process requires flexible expander operation regarding mass flow and differential pressure. To meet this requirement the new and advanced LNG expander is designed as a hybrid machine combining a radial inflow reaction turbine and an axial propeller turbine with variable blade pitch.

Figure 5 shows a cross section of the Exducer Turbine, the third generation of LNG expanders. Unlike the first and second generation the main LNG flow direction is upward and against the gravity. LNG is slightly compressible and the convective forces support an upward flow direction. Equal to the second generation of LNG expanders, the entire Exducer Turbine is completely submerged in LNG to eliminate dynamic mechanical seals and couplings.

In this example, the radial inflow reaction turbine with fixed geometry guide vanes consists of three stages to expand the majority of the fluid pressure and to generate the equivalent amount of electrical power. The exducer stage, a uniquely designed axial propeller turbine, is mounted on the same shaft directly above the reaction turbine. The blade pitch of the exducer rotor is continuously adjustable and allows operation of the expander at the Best Efficiency Point within a certain range of varying mass flows and differential pressures.

Figure 6 shows a cross section of the exducer stage. The particular design of the exducer rotor consists of a concentric hemispherical shroud and hub. The shape of the blades corresponds to the sector of the circular cross-section of shroud and hub and the blade rotational axis intersects with the centre of the concentric hemispherical shroud and hub. This hemispherical design is the only possible geometrical solution for adjustable blades able to seal shroud and hub for all angular blade positions, respectively.

The adjustment range of the exducer blades is between horizontal and vertical corresponding to completely closed and completely open blade positions. The blades are fitted with guide vanes to direct the flow through the hemispherical exducer. The guide vanes are shaped as concentric spherical shells.

Figure 8 shows the typical performance of Exducer Turbine Expanders as a function of differential head and flow. In the case of no-load operation (8) with a de-energized generator, the expander is operating along one of the no-load characteristic curves N_X. The position of these curves N_X depends on the angular position X of the exducer blades. If the blade position for the rated case generates the no-load characteristic curve N_0, then by closing the exducer blades, the no-load characteristic curve moves to the position N_1 towards higher differential pressures. By opening the exducer blades, the no-load characteristic curve moves to the position N_2 towards lower differential pressures.

The corresponding orifice curve for the rated case is L_0. For the closing and opening blade positions the corresponding curves are L_1 and L_2. If the generator is energized and produces electrical power at the fixed rotational speed, then the exducer turbine expander operates for the rated case along the performance curve T_0 to the right of N_0 and above L_0. If the exducer blade position is closing or opening, the performance curves T_1 or T_2 to the right of N_1 or N_2 and above L_1 or L_2 determine the expander operation.

All of the performance curves T_X for different blade positions X intersect at one point located at the coordinate axis for the differential head. The location of this point is determined by the rotational speed of the expander. With an increasing flow rate, the performance range is increasing. For any given values of flow and differential head within this range, there is a certain exducer blade position that generates the correct performance curve to meet these values.

6 TWO-PHASE LNG EXPANDERS

The thermodynamics of gas liquefaction processes indicate significant advantages if the expansion of liquefied gas across an expansion machine is partially or completely carried out into the vapour phase. The design of the Exducer Turbine Expander with the added feature of variable speed is capable to expand liquefied gas into the vapour phase. The vaporization takes place within the exducer stage and is controlled by the angular position of the exducer blades and the rotational speed.

Figure 9 shows the typical performance of the two-phase exducer turbine. Let T_0 be the performance curve of the two-phase expander for a given rotational speed and given blade position. By increasing the rotational speed the performance curve moves to T_1 or alternatively by changing the angular blade position the performance curve moves to T_{01}.

For any given flow within a certain range the two-phase expander is able to meet the correct overall differential pressure and the correct intermediate differential pressure to vaporize the liquefied gas to the specified degree by applying the two control parameters of the expander: variable speed and variable blade pitch. Figure 10 shows the performance field of two-phase exducer turbines operating with variable speed and variable blade pitch.

The liquid expansion is performed across the radial inflow reaction turbine stages and the vaporization is performed across the exducer stage. To meet this requirement the saturation point of the liquefied gas, which is determined by the thermophysical properties, has to be located between the last stage of the reaction turbine and the exducer turbine. The two control parameters of the two-phase expander, the variable speed and the variable exducer blade pitch enable the two-phase expander to adjust the saturation point across the stages of the turbine expander.

7 ACKNOWLEDGEMENT

The authors wish to acknowledge the generous assistance received from Mr. Norrazak Haji Ismail, General Manager, Technical Division, Petronas, Malaysia LNG Tiga and from Mr. John Edward Bol (PMT), MLNG Tiga Plant Project, Main Site Office, Bintulu, Sarawak, Malaysia, regarding liquid expanders at MLNG Dua and Tiga.

8 REFERENCES

(1) Kimmel H.E. "Variable Speed Turbine Generators in LNG Liquefaction Plants" Proceedings of the GASTECH '96, Vienna, Austria, December 1996.
(2) Verkoelen J. "Initial Experience with LNG/MCR Expanders in MLNG-Dua" Proceedings of the GASTECH '96, Vienna, Austria, December 1996.
(3) Baines N.C., Oliphant K.N., Kimmel H.E., Habets G.L.G.M. "CFD Analysis and Test of a Fluid Machine Operating as a Pump and Turbine" IMechE Seminar Publication, CFD in Fluid Machinery, 15 Oct. 1998, London, U.K., ISSN 1357-9193, ISBN 1 86058 165 X
(4) Habets G.L.G.M., Kimmel H.E. "Economics of Cryogenic Turbine Expanders" The International Journal of Hydrocarbon Engineering, December/January 1998/99, Palladian Publications, U.K.
(5) Habets G.L.G.M., Kimmel H.E. "Development of a Hydraulic Turbine in Liquefied Natural Gas" IMechE Conference Transactions, Seventh European Congress on Fluid Machinery for the Oil, Petrochemical, and Related Industries, 15-16 April 1999, The Hague, The Netherlands, ISSN 1356-1448, ISBN 1 86058 217 6, London, U.K.
(6) Song M.C.K., Kimmel H.E. "Cooling Cycle Expanders Improve LNG Liquefaction Process" Third Joint China/USA Chemical Engineering Conference, September 2000, Beijing, China
(7) Van den Handel R.J.A.N., Kimmel H.E. "A New Generation of Liquid Expanders in Operation at Oman LNG" Proceedings of the GASTECH 2000, Houston, Texas, USA, November 2000.
(8) Hylton E.H., Kimmel H.E. "Upgrading Existing LNG Plants Using Exducer Turbine Expanders" GASEX 2002 Brunei Darussalam, Conference Papers, 27-30 May 2002

Figure 1: First generation LNG expander with air-cooled generator, variable guide vanes and rotating shaft seal.

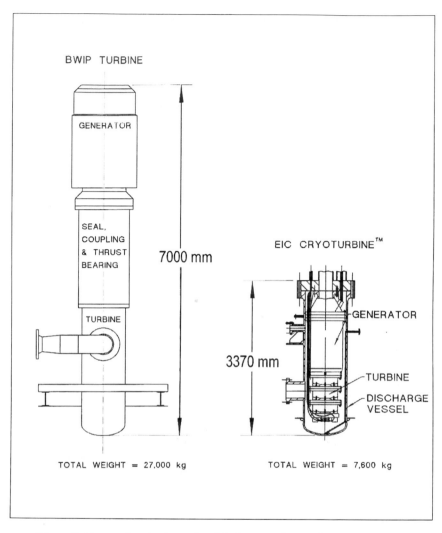

Figure 2: Comparison in size and weight between first and second generation LNG expanders.

C603/017/2003 © IMechE 2003

Figure 3: Second-generation seal-less LNG expander with submerged generator and variable speed control

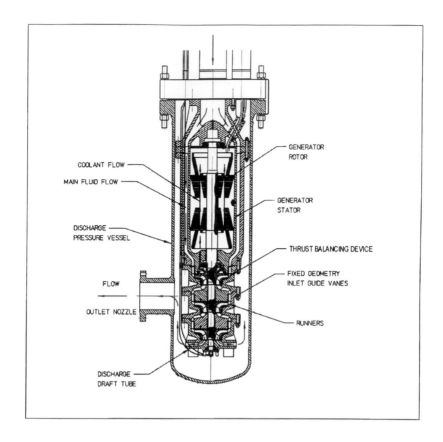

COOLANT FLOW

MAIN FLUID FLOW

DISCHARGE
PRESSURE VESSEL

FLOW

OUTLET NOZZLE

DISCHARGE
DRAFT TUBE

GENERATOR
ROTOR

GENERATOR
STATOR

THRUST BALANCING DEVICE

FIXED GEOMETRY
INLET GUIDE VANES

RUNNERS

Figure 4: Basic design of second generation submerged LNG expanders.

C603/017/2003 © IMechE 2003

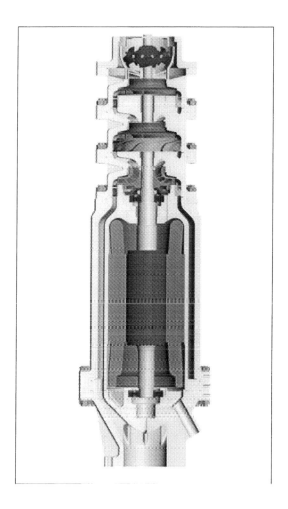

Figure 5: Third generation LNG expander with constant speed and variable blade pitch exducer stage.

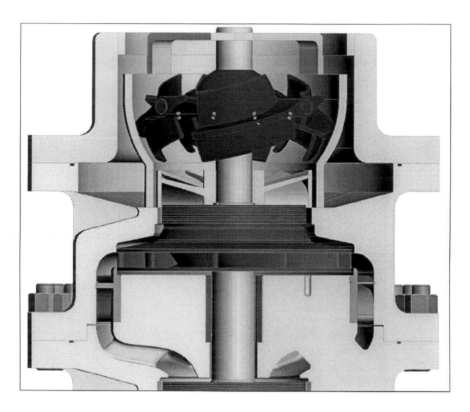

Figure 6: Cross section showing the combination of radial turbine stage with exducer turbine stage.

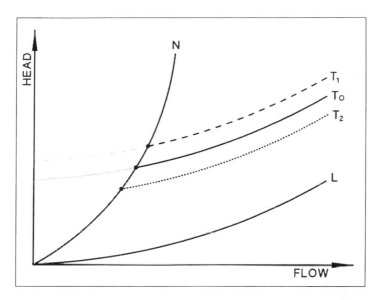

Figure 7: Typical Performance for Variable Speed LNG Expanders

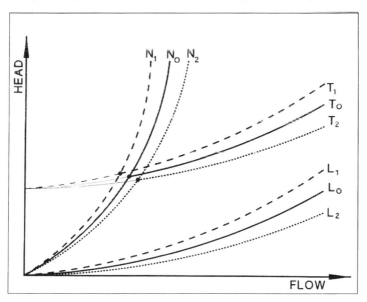

Figure 8: Typical Performance for Exducer Turbine Expanders

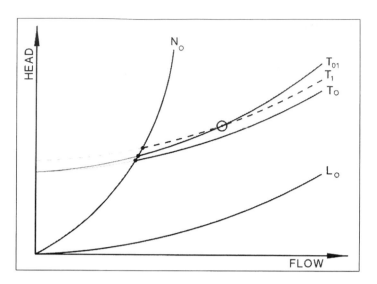

Figure 9: Typical performance of two-phase exducer turbine.

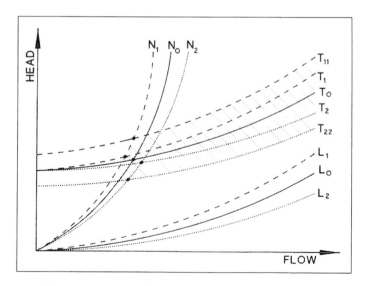

Figure 10: Performance field of two-phase exducer turbine.

C603/017/2003 © IMechE 2003

C603/024/2003

Design of large-scale air separation turbomachinery units

H VOSS
Man Turbomaschinene AG, Oberhausen, Germany

SYNOPSIS

Design of large scale air separation turbomachinery trains in the range of 50 MW to 80 MW are in increasing demand for use in air separation applications integrated with large scale process /industrial plant, including the following sectors :

-   Enhanced Oil Recovery (EOR) : nitrogen injection schemes
-   Integrated Combined Cycle (IGCC) power schemes employing coal and refinery residues
-   Bulk methanol production schemes
-   Gas to Liquids (GTL) projects for stranded gas reserves

In many instances steam is generated within the main process plant and the air separation plant power demand is integrated within the overall energy balance to optimise plant efficiency.

The turbomachinery designer is faced with selecting machine designs which must handle large volume flows (air inlet volumes up to 1,300,000 m³/h & steam exhaust volumes up to 1,600,000 m3/h). Some limitations on train sizing are imposed by the maximum available shell and tube heat exchangers, the size of nozzles needed to introduce compressor intercooling and if necessary, steam reheat connections to limit the exhaust wetness.

The paper describes how the turbomachinery design engineer, from both a compressor and steam turbine aspect, solves all the demands of the process energy balance, the physical size limitations and the dynamics of the coupled plant systems, into a machine train design that fully satisfies the demands from the major oil companies for referenced equipment.

1 INTRODUCTION

The trends towards optimisation of large turbomachinery in terms of efficiency versus cost plays an important role in the evaluation and selection of compressor type for a given project.

Generally, evaluations are carried out not only on the performance of each component of the rotating machinery, but also on the compactness, the reliability and availability of the equipment. Since trains in production operate continuously they are subjected to an agreed maintenance program. Therefore, from the user's as well as the manufacturer's point of view a common objective has emerged to select and use optimised machinery for a project. These ideas have brought forth new process designs as well as major development programs that have been undertaken by manufacturers to upgrade the design and selection of machinery. Such studies were conducted to apply major efforts towards the development of high-performance turbomachinery.

The complex design processes required for such large turbomachinery trains led customers to obtain such technology from a single source. After integration of the turbomachinery activities from SULZER, the MAN TURBO Group comprises today three core companies with a leading position in the air separation machinery market and other key process industries.

2 AIR SEPARATION PLANTS

In the air separation unit (ASU) refrigeration takes place primarily by expanding almost isothermally compressed air. The compression of air in the range of 6 to 10 bar is performed by multistage turbocompressors with intermediate- and aftercooling. After drying the air and CO_2 removal, modern oxygen plant designs (based on a pumped liquid cycle) feature a dry air booster which compresses part of the air flow to a pressure roughly twice that of the final gaseous oxygen. In gasification processes where oxygen injection is required in the range 70 – 80 bar, oxygen compressors are still required.

Compressors associated with air separation plant are
- Main Air Compressor   (MAC)
- Booster Air Compressor   (BAC)
- Oxygen Compressor
- Nitrogen Compressor

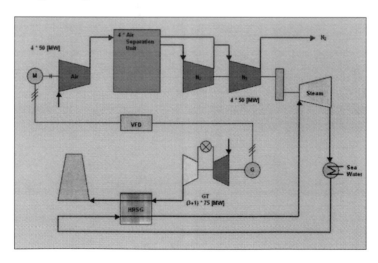

Fig. 1: ASU Plant Schematic

C603/024/2003 © IMechE 2003

2.1 Enhanced oil recovery

Fig. 1 shows a schematic of the world's largest nitrogen production plant in the Gulf of Mexico. The diminishing well pressure of the crude oil is to be increased by injecting about 1.3 million Nm³/h nitrogen at 121 bar. For this project, MAN TURBO was awarded the order to supply the air compressors for the air separation plant and the nitrogen compressors including their steam turbine drivers. The total installed driver power for compression exceeds 400 MW.

The four gasturbine units – gas turbines, generators and heat recovery steam generators (HRSG) – deliver electricity and heat for four 50MW air compressors with direct motor drive and the four nitrogen compressor sets with 55MW condensing steam turbines.

2.2 Gasifiers for syngas production

Conversion of coal, orimulsion, petroleum coke & other refinery bottoms into syngas in a gasification unit requires the introduction of oxygen at pressure into the gasifier. The syngas produced can be used as feedstock for a petrochemical process or as fuelgas for power generation (IGCC).

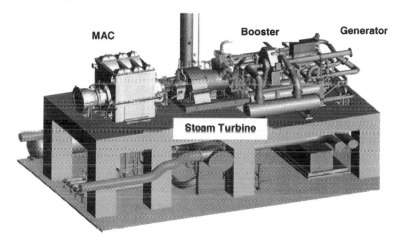

Fig. 2: Typical Machinery Train for a Methanol Plant

In some installations the ASU turbomachinery train is integrated into the process plant. Such an example is a train currently in manufacture for a large merchant methanol facility in Trinidad (Fig. 2). The unit is driven by a 55 MW double-ended steam turbine and the plant energy balance is such that a high degree of intercooling is required for the air compression train. The MAC rated at 500,000 m³/h is a radial compressor with integral cooling. The BAC is a multistage gear type compressor with external coolers.

2.2 Gas to Liquids

As an alternative to long term Liquid Natural Gas (LNG) contracts, owners of stranded gas reserves are looking ever increasingly to "gas to liquids" processes (GTL) as a means of selling low-sulphur fuels to consumers on the spot market.

The GTL process can be split into three sections :

     i)   production of syngas from natural gas by combining oxygen with carbon

     ii)   conversion of syngas to synthetic crude

     iii)   refining synthetic crude into synthetic fuels

Many of the GTL processes on the market employ the Fischer Tropsch process to convert syngas to paraffinic hydrocarbons (synthetic crude). This is an exothermic process which generates low pressure steam; high quality steam is produced in addition during syngas production. Because of the surplus energy available from the process, the steam generated is used to drive the ASU compression train; this means that instead of looking for the highest ASU turbomachinery train efficiency, the compressor intercooling can be optimised to obtain an energy balance across the whole site complex. A typical main air compressor for a GTL plant is shown in Fig. 3.

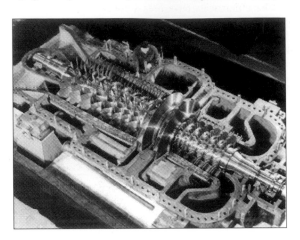

Fig. 3: Typical Main Air Compressor

In these plants very large turbomachinery trains are required to deliver suction volumes of between 600,000 – 1,300,000 m³/h to the ASU main air compressor.

3 MAIN AIR COMPRESSOR DESIGN

The air separation market itself is highly cost oriented and competitive. For processes which need oxygen for the production of synthesis gas, large air compressors up to about 700,000 m³/h are used. Important requirements for the compressor selection are
- High Priority on Efficiency
- Low Investment Cost.
- High Reliability

In many applications the production output of the process and the gas supply are coupled so that the compressor has to adapt to 65 and 110 % flow at constant discharge pressure. In order to meet these variations of flow, industrial compressors are tailored according to the customers specified data. To keep fabrication costs low and the design process reliable the compressors are built out of a modular system of standard blades, impellers and casing sizes.

3.1 Axial or radial
Axial compressors were developed at the start of the 1950s and have since been introduced very successfully into various market segments of the process industry. While their primary use initially was in blast furnace applications, the emphasis has shifted over time

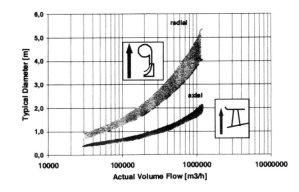

Fig. 4: Axial versus Radial Casing Diameter

increasingly to the chemical and petrochemical sector, and also to the " air separation plant" segment of the market.

Over the last ten years, centrifugal impeller efficiencies have been improved by using CFD-design codes. The efficiency advantage of axial stages vs. radial stages tends to reduce from 5 to 2 % for large volume applications; although for high flow applications only axial designs can be considered from a practical standpoint. These two types of compressor configurations differ very strongly in their frontal area (Fig. 4); in the case of 700,000 m³/h suction volume flow, the axial front stage diameter is 1,550 mm whereas the outer diameter of radial stage volute exceeds 3,200 mm for an impeller diameter of 2,000 mm.

For high flow compressor applications, the benefits of the smaller dimensioned axial flow compressor front-end are combined with a number of centrifugal stages; these final radial stages allow the level of intercooling necessary to achieve an energy balance across the plant.

For high air flowrates, the maximum intercooler sizes available on the market have an influence on the pressure at which the first stage of intercooling is introduced in the compressor. The resulting rotor configuration gives a good balance between the dimensions of

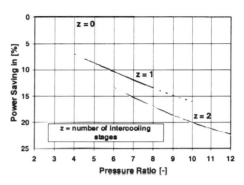

Fig. 5: Influence of Intercooling on Power

the axial and centrifugal sections. The influence of intercooling on shaft power is shown in Fig. 5.

3.2 Blading Design

Improving the machine performance and reducing machinery dimensions is a prime target for every supplier. At present, development work is focused on the following areas:

- higher stage pressure rise
- increased efficiency
- maximise flow for a given casing size and
- maximise the operating range from surge to choke.

The requirement for a typical industrial axial flow compressor is to allow long term operation across the entire operating range, i.e between surge and choke (1). A wide operating range is achieved by keeping the inlet Mach number at the blade tip in the subsonic region. The related circumferential speed therefore is also limited which results in a relative low stage pressure ratio.

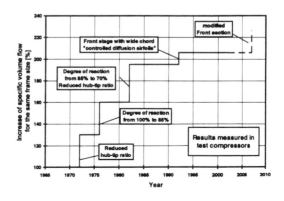

Fig. 6: Flow Increase during Front Stage Development

However, to obtain a reduction in machine size, the flow velocities through the front stages have to be increased to the maximum allowable limit. This results, during conceptual design, in a compromise between operating range, size and number of stages.

Great efforts have been taken especially in the front stage design (2, 3) which made great progress in the past years. During a development period of about 20 years the volume flow could be doubled for the same size of axial compressor (Fig. 6). The operating range remained unchanged. This was achieved by different development steps which concentrated on reduced hub-tip ratio, increasing speed, and by reducing the degree of reaction of the first few axial stages.

Compared with the rear stages in a multistage compressor the front stages are developed to perform at high positive incidences and high specific volume flows.

The maximum volume flow for an axial compressor cascade cannot be achieved with an axial direction of the inlet velocity, a certain preswirl is needed to optimise the suction capability of the front stages (4) which can be defined as Ω:

$$\Omega = \frac{\overset{\bullet}{m}\ \omega^2}{\rho_0\ a_0^3}$$

with $\overset{\bullet}{m}$ as mass flow, ω as angular velocity, ρ_0 as inlet density and a_0 as velocity of sound. Fig. 7 shows the suction ability - corrected by ν as the hub/tip ratio - in relation to the inlet swirl angle α_1. In

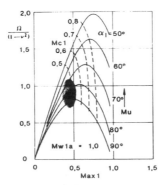
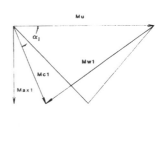

Fig. 7: Aerodynamic Flow Optimisation

the chart is also marked the typical front stage design area correlated for a blade tip Mach number of 0.8 instead of 1.0. The diagram seems to present much higher flow capabilities for the front stages but a further increase of Mach number levels narrows the allowable incidence variation for a wide range of operation.

Modern controlled diffusion design techniques were applied to develop a new front stage with a wide chord blade (Fig. 8) which has been proven in several customer´s compressors. The advantages of such a low aspect ratio are higher pressure ratio and better operating range, coupled with lower mechanical blade stresses. During surge the alternating stresses at the blade root could be halved.

Fig. 8: Front Stage with Wide Cord Blade

C603/024/2003 © IMechE 2003

3.3 SCALING EFFECTS

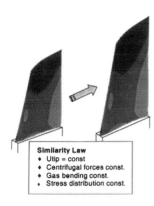

Similarity Law
- Utip = const
- Centrifugal forces const.
- Gas bending const.
- Stress distribution const.

Fig. 9:Blade Development

3.3.1 Blading

Thermodynamics and mechanics use laws of similarity to adapt to different sizes of geometry. Scaling blades or impellers of compressors means that geometrical relations are kept constant to retain special aerodynamic and mechanic features (Fig. 9). Design criteria based on aerodynamic similarity are well accepted and increasing the size of components judged using such methods is not considered to be a problem, whereas confidence in mechanical proofs and production safety is more conservative. It may be understood as a safety rule. Any mechanical failures in turbomachinery lead often to big damages and production losses. Therefore references are typically valid only for small steps in size increase. Likewise, the adoption of mechanical similarity rules, results in the same static stress loading of turbomachinery blades. What may be different is the vibrational situation because of resonance points (Campbell diagram) and changed alternating stress levels (Goodman diagram). Manufacturers of turbomachinery perform different proofs for blade mechanics, which allow moderate blade length increase under similarity laws and incorporate a certain factor of safety.

3.3.2 Compressor casing design

Compressor casings can be manufactured by two different production technologies: casting and welding. Both technologies have been applied successfully by MAN TURBO. The decision which technology will be used depends on several factors, such as number of identical casings, cost of patterns, delivery time etc. For multiple machine trains, as delivered to Cantarell, the cast casing design certainly gives the lowest unit cost.

Welding techniques however have improved over the past years and complicated structures like casings with several nozzles can be fabricated by the use of robot welding. For one off machines fabricated designs are often employed to avoid pattern costs.

3.4 Axial air compressor

Since 1977 MAN has supplied a total of 14 identical axial flow compressors, each with 36 MW (48275 HP) drive power, to the world's largest oxygen complex in Sasolburg, South Africa. These compressors have today successfully clocked up around 2 million hours under severe operating conditions.

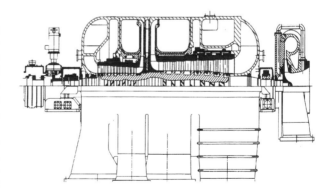

Fig. 10: Intercooled Axial-Flow Compressor

The axial flow compressors used for the Cantarell nitrogen injection scheme in Mexico are similar to those for the Secunda RSA (Fig. 10) installation. Each of the four machines operates at over 50 MW (67050 HP) with suction volumes approaching 600,000 m3/hr (353100 acfm) depending on plant demand and ambient conditions. The power required is supplied by a synchronous motor directly driving the compressor at 3,600 rpm and started by a soft starter. Its rated drive output is 52 MW (69733 HP).

The air is taken in via a radial inlet casing and following compression in seven axial stages and a radial end stage in the low-pressure section, is conveyed to the intercooler located beneath the compressor (Figure 11). Following further compression in the intermediate-pressure section (six axial stages and radial end stage) and a further stage of intercooling, the final compression is carried out in the overhung-mounted radial stage with an axial intake.

The wide range of compressor performance data required can be delivered with just four synchronously adjustable rows of stator blades at the inlet to the low-pressure section. The reasons for the superb partial load characteristics lie in the moderate aerodynamic loading of the blading of the low- and intermediate-pressure section together with the stage characteristics of the HP stage.

The dimensions of the axial flow

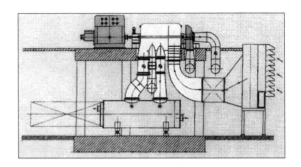

Fig. 11: MAC Compressor Arrangement

compressor can be characterised best by the distance between bearings of around 6 m, the outer diameter of the largest centrifugal impeller of 1.64 m and the transportation weight of roughly 180 tonnes.

4 BOOSTER AIR COMPRESSOR DESIGN

4.1 Compressor type
It is common to use integrally geared centrifugal compressor designs for ASU air booster compressor duties, especially where high energy costs are applied; the trains supplied to the methanol plants in Singapore and Trinidad employ this style of machine.

For GTL applications the tendency for the oil company (end user) is to prefer single shaft API617 designs. The power density for GTL air booster applications is also in the range of 4 to 5 MW per impeller, where again single shaft designs are the more conservative option.

4.2 Booster compressor construction
The booster compressor construction used for GTL applications, is a single shaft barrel design with the impellers arranged in three stage groups to allow external intercooling; refer to Fig. 12. This type of compressor has been successfully employed as the high pressure casing of the Cantarell nitrogen compressor trains; whilst of a similar size for a 3500 tpd ASU, the boost pressures are around 70 – 80 bar, rather than the high injection pressure needed for EOR.

C603/024/2003 © IMechE 2003

For the ASU Cantarell, each of the four nitrogen centrifugal compressor trains (designed to API Standard) operates with shaft powers up to 48 MW depending on the well demand profile and pressures up to 121 bara (1755 psia). The high-pressure nitrogen compressor is provided with two stages of intercooling and contains three stage groups, each comprising two impellers with an outer diameter of 660 mm. The casing of cast steel is designed as a barrel with a single cover. This compressor features alternating vaned diffusers and volutes with a circular cross-section to optimise

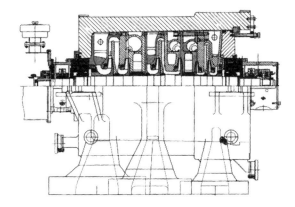

Fig. 12: Typical Booster Compressor

efficiency in the transfer of compressed nitrogen to the coolers located beneath the machine. The initial stages of the three HP stage groups have vaned diffusers. The HP compressors are constructed without an internal casing; in place of this, the internals are connected via special tie-bolts to form the axially removable cartridge. The compressor was subjected a full load test with a FT8 gas turbine package as a driver (5).

5.0 STEAM TURBINE DRIVER

5.1 Turbine design
The industrial range of steam turbines built by MAN TURBO, are based on a modular design matrix. As the casing elements have been standardised, there also are various series of standard nozzles, stator and rotor blades available. The blades used in the high and medium pressure parts are completely milled with integral shrouds. Both condensing and back-pressure designs are available, with controlled and uncontrolled admissions/extractions and bleed connections for live steam conditions up to 130 bar, 570°C. Power outputs are ranging up to 120 MW in a single casing.
Features of the turbine design include :-
-      centre-line mounted casing
-      separate bearing pedestals
-      fabricated exhaust casing welded to main turbine housing (radial and axial)
-      variable geometry extraction control
-      transonic exhaust stage

5.2 Turbine selection for combined MAC/BAC drive
As the main air (MAC) and booster air (BAC) compressors are inextricably linked by the air separation process, and where the selected driver is a steam turbine (Figure 13), it makes sense to integrate the various turbomachine casings into a common train.
With turbine ratings in the order of 80 MW for a 3500tpd ASU, a good speed match can be found between the axial/radial compressor (MAC) and the steam turbine driver. The turbine has a double-ended arrangement, which allows the booster to be driven from the second shaft end via a parallel shaft gearbox.

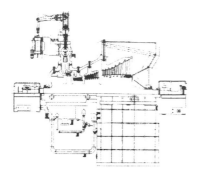
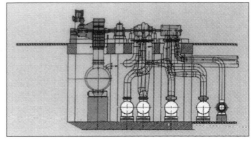

Fig. 13: Industrial Steam Turbine   Fig . 14: Nitrogen Radial Compressor Arrangement

5.3 Similar units
Similar condensing steam turbine designs have been supplied to drive the four Cantarell nitrogen compressor trains, each with a rating of 55 MW. Details of the steam turbine - compressor arrangement are shown in Figure 14.

The steam turbine for the Trinidad train, is rated at around 57MW, is similar in size and arrangement to the double-ended units selected for the 3500 tpd ASU's for GTL service.

6 DYNAMIC SIMULATION

Unlike steady-state simulation, dynamic simulation permits modeling process transients. This provides a deeper understanding of the process, which usually results in improved performance.

In an air separation plant design the dynamic simulation is used to satisfy the customer's strict set of requirements for dynamic response. The design issues involved control strategy development, process equipment sizing considering dynamic performance, initial controller tunings and plant performance assessment.

The benefits from using dynamic simulations are developments of safe start-up and shutdown

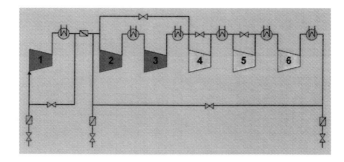

Fig. 15: Nitrogen Compressor Loop

strategies.

MAN began at an very early stage to analyse the behaviour of turbomachinery systems under the aforementioned operating conditions (6). They accumulated numerous experiences on the test facility and in the field to validate the theoretical models.

In the dynamic simulation, the thermodynamic performance curves of the compressors, the

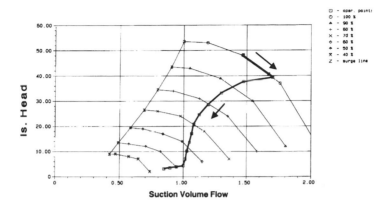

Fig. 16: Result of Dynamic Simulation

characteristics of the drive units, the mass moments of inertia of all rotating components, the storage volumes of the gas coolers and the gas-carrying pipelines and the corresponding valves and fittings as related to one another by means of a system of differential equations, which are solved digitally. The attributes of the MAN turbolog DSP controller is that it can be fully integrated into the investigation without the need to model the controller itself. As a result of this simulation, all the important time-dependent parameters, such as speeds, pressures, volume flows etc. in the transient behaviour of the installation are available for further evaluation. By varying elements of the plant, e.g. the arrangement and attributes of valves and fittings, this behaviour can be influenced in a targeted manner even during the plant design stage.

One example of using this simulation system is shown in Figure 15, showing the system design of the Cantarell nitrogen compressors. To avoid „surging" of individual stages during trip or stage groups in transient processes, recycle valves are provided for certain stage groups, which open automatically when required. Figure 16 shows the result of such a simulation into the transient effects of a shutdown process – in this case by means of turbine tripping – and which can be traced in the performance map of the 6th stage group of the nitrogen compressor.

7 REFERENCES

(1) Voss, H., „Axial-Flow Compressors Between Surge and Choke", IMechE, 1999, C542/051.

(2) Eisenberg, B., „Development of a New Front Stage for an Industrial Axial Flow Compressor", Transactions of the ASME, 604, Vol. 116, Oct. 1994.

(3) Eisenberg, B., Thomas, P., and Turanskyj, L., „Effects of Rotating Stall and Surge on the First Rotor Blade Row in a Six Stage Industrial Axial Flow Compressor", VDI-Berichte 1208, 1995.

(4) Voss, H., „Beanspruchungsanalyse der hinteren Stufen von Industrie-Axialverdichtern", Diss. Uni. d. BW Hamburg, 1994.

(5) Aschenbruck, E., Blessing, R., Turanskyj, L. "FT8-55 Mechanical Drive Aeroderivative Gas Turbine: Design of Power Turbine and Full-Load-Test Results", ASME Paper 94-GT-343, 1994.

(6) Voss, H., „Mathematical Simulation of Turbocompressors in Chemical Processes", IMechE, C55/81, 1981.

C603/031/2003

The MS 5002E – a new two-shaft, high-efficiency, heavy-duty, gas turbine for Oil&Gas applications

L AURELIO, L TOGNARELLI, and P PECCHI
GE Oil&Gas – Nuovo Pignone SpA, Florence, Italy

1. ABSTRACT

A new version of the very successful MS5002 (GE Frame 5 two-shaft) has been developed considering both mechanical drive and Power generation markets, with the aim to satisfy the most recent Customer requirements in terms of high efficiency and therefore fuel consumption and environmental impact.

Power class is 30 MW, pressure ratio is 17:1, simple cycle efficiency is over 36% and combined cycle efficiency approximately 51%.

The new model retains features that contributed to the success of its predecessors:
- full heavy-duty concept for on-site maintenance
- moderate firing temperature (compared with state of the art) for highest reliability
- two-shaft design with free power turbine for mechanical drive use
- high heat recovery capability.

Achievement of high cycle efficiency with low firing temperature is possible thanks to the advanced tools used for the definition, design and optimization of airfoils, clearances, leakages and distribution of cooling flows.

The dry-low-emissions combustion system design is derived from the GEPS DLN2, a single stage dual mode combustor that can operate with both gaseous and liquid fuel.

A thorough testing program, including the full-scale test of the axial compressor and a full load prototype test, has been established to support development and to validate the design.

2. INTRODUCTION

The gas turbine market is increasingly oriented toward high efficiency and low emissions, due to the ever higher fuel cost and to stricter environmental regulations.

This is becoming true also for the mechanical drive market, where in the past availability and reliability used to be almost the only required features.

For this reason GE Oil&Gas decided to develop a new version of the well-known MS5002 (fig.1). The MS5002E is targeting 36.4% simple cycle shaft efficiency, 25 PPM NOx emissions.

In order to retain availability and reliability comparable with the previous models, the design firing temperature was set at a conservative value.

Power will be in the range of 30+ MW, comparable with that of the MS5002D.
The MS5002E is not intended to replace completely the previous models. The MS5002C and D will remain in production, in order to provide a high number of product options able to cover the most diverse requirements.

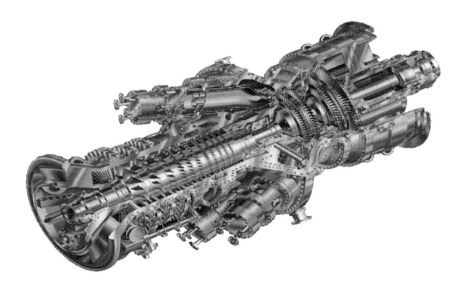

Fig. 1 – MS5002E breakout section

3. NOMENCLATURE

DFSS – Design for Six Sigma
DLN – Dry Low NOx
DOE – Design of experiments
DS – Directionally solidified
GE – General Electric
GEPS – General Electric Power Systems
HP – High Pressure
IGV – Inlet Guide Vanes
LP – Low Pressure
NPI – New Product Introduction

4. DEVELOPMENT PROCESS

The machine was designed following the NPI (New Product Introduction) process. This is a disciplined GE approach, requiring the completion of a number of sequenced steps, each subject to review and approval by the Company leadership team.

A Design for quality methodology (Design for Six Sigma – DFSS) is embedded in the NPI process. DFFS is a quality initiative used at GE to develop new products; it permits to design using a structured process that addresses the customer requirements (Critical To Quality – CTQs).

Engineering design reviews must be held after any major design step. Reviewers are members of the Chief Engineers Office, so that the experience of the whole Company is factored into the design.

5. TECHNICAL FEATURES

5.1. Compressor
The 11-stage compressor (fig. 2) is directly scaled-up from that of the existing GE10 gas turbine, on which development work was performed to achieve top performance with a minimum number of stages [1]. Pressure ratio is 17:1, airflow is 96 Kg/sec. The nominal operating speed is 7455 rpm. The inlet guide vanes (IGV) and the first and second stage stator blades are variable for compressor speed control and to avoid stall during start-up at reduced operating speeds. The variable stages are hydraulically actuated through master levers and rotating rings, the same method as used on the GE10 gas turbine.

Two bleed ports are present along the flowpath. The 4th stage bleed is used for LP turbine wheels cooling and bearings sealing.

The 7th stage bleed is used for cooling and for surge control during start-up/shut-down. Conventional airfoils were replaced by "custom tailored" airfoils, each one individually designed to optimize Mach number distributions and minimize shock and profile losses.

Fig. 2 - The FR5 2E Compressor Rotor

A feature of this advanced design technique is the special 3D airfoil stacking method, whose main purpose is to minimize secondary flow effects and boundary layer flows that lead to local flow separations and increased losses.

In order to maintain the same aero-mechanical behavior, the rotor has the same structural design as the GE10: one forward stub shaft, six discs and five spacers, one aft stub shaft, all packed together by 26 tie bolts. Materials are Cr-Mo-V steel for the shaft and bolts. Blade materials are 17-4 PH (1-2-3 rotor stages) and AISI 422.

The compressor casings are horizontally split for on-site maintenance, as in the existing FR5 design.

The air inlet casing supports the Bearing #1, a combined tilting-pad journal and thrust bearing. Casing materials are cast iron for inlet case, nodular cast iron for the intermediate case, cast steel for the compressor discharge case.

5.2. Combustion system

The combustion system is of the can-annular, reverse flow type, with six cans mounted on the compressor discharge case. It is derived from the GEPS DLN2 installed on "F" Class machines and operates with gaseous fuel. Future development will include liquid fuel capability.

C603/031/2003 © IMechE 2003

In each combustion can there are a 4+1 fuel nozzles, each of them containing a premixing tube, where fuel gas and air mix together before the primary burning zone, and a central body, with a diffusion fuel gas circuit.

Since in the MS5002E the firing temperature is lower than in the "F" turbines, the resulting combustor design is particularly conservative in all respects, including components life and low emissions capability.

The combustor operates in diffusion mode at low loads (less than 25%), piloted-premix mode from 25 to 50% and in premixed mode at higher loads (more than 50%) with a 25 PPM NOx initial target. The fuel gas delivery system is provided with multiple gas control valves to distribute the gas fuel flow to the different gas circuits. Accurate split is required during premix operation to ensure both low emissions and low combustion dynamics.

The tests were performed preliminarily at GE Corporate Research and Development Center in Schenectady (NY), to assess the combustor behavior over the entire range of operation. Design validation tests have been also performed at the full-scale test rig at GE Power System plant in Greenville (SC). During the production phase, the test rig will be available to test the combustor performance with different fuels as required by specific customer applications.

5.3. HP Turbine

To maximize reliability, a low firing temperature was chosen. To get high simple cycle efficiency, state of the art 3D aero and advanced airfoil cooling systems (similar to those extensively used in aircraft engines) were implemented.

The axial flow, two-stage reaction type HP Turbine, with air-cooled nozzles and buckets was designed to obtain high efficiency over a wide power range. The Turbine was aerodynamically designed using advanced three-dimensional computational fluid dynamics techniques.
It consists of two turbine wheels, first and second stage turbine nozzle assemblies, and turbine casings.

The Stage 1 nozzle is made of un-coated FSX 414, with two airfoils for each sector and it is mounted in a retaining ring at the outer band. The Stage 1 nozzle assembly is air cooled (convection and film) by compressor discharge air flowing through each vane. Internally, the vane is divided in two cavities that house the impingement inserts. The air flows in opposite directions in the two inserts. In the trailing edge insert air enters from the top, while in the leading edge insert air enters from the bottom. Air flowing into the forward cavity is discharged through holes in the leading edge and on each side close to the leading edge. Air flowing into the aft cavity is discharged through additional film holes and trailing edge slots provided with turbulence promoters.

Stage 1 buckets are made of coated GT111 DS material. The buckets have three tangential dovetails. The shank has two "angel wings" to prevent hot gas ingestion in the rotating cavities. A sealing pin is inserted in the shank and leaf seals are used to reduce gas leakage (fig. 3)

The buckets are cooled by air that flows through the dovetails and blade shanks. The internal cavities provide convection and external film cooling by passing the cooling flow through the internal serpentine circuits provided with turbulators, with a part discharged through the leading and trailing edges for film cooling.

Stage 2 nozzle is made of GTD 222 with Aluminide coating. The nozzles are cantilevered from hooks on their outer band, fitting into slots in the support structure. Each sector includes three airfoils each one having a single cavity, provided with a welded insert on the outer platform for impingement cooling.

The Stage 2 bucket is made of coated GTD111 Equiax. Air enters from the top and is discharged through the outlet insert tube and the TE holes with turbulators.

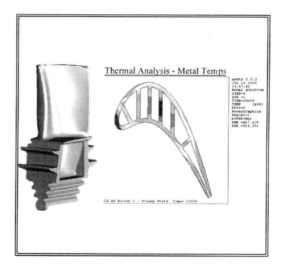

Fig 3 – 1st stage bucket with angel wings on shank and a sample of thermal analysis

5.4. LP Turbine

The power turbine is the same module of the LM2500+HSPT, which was designed using the heavy-duty concept, even if coupled to an aeroderivative gas generator [2]. The mechanical structure is the same, however, flowpath profile and airfoils were re-designed because of the higher required airflow.

 C603/031/2003 © IMechE 2003

The power turbine is a two-stage design aerodynamically matched to the gas generator. The hot gases leaving the gas generator are directed into the power turbine inlet liner. The exhaust gases leave the turbine through the diffuser, and then are turned 90° by the exhaust ductFirst and second stage rotor blades are provided with interlocking tip shrouds to minimize vibration levels, shanks to reduce the heat flow towards the disks, and are retained in the disks by dovetails. Pins seal the inner space between the platforms and the shanks. To prevent hot gas ingestion in the rotating cavities between the disks, the blades contain two "angel wing" seals at the inlet side of the shank and one at the outlet side.

The power turbine stator consists of three turbine casings. The 1st stage turbine casing is a shell, which contains the shrouds ring. Each shroud, that acts as a casing heat shield, has a honeycomb surface, which provides a close clearance seal between the rotating blades and the shroud itself. The 2nd stage casing has the same configuration as the 1st stage casing. The nozzles of both stages consist of segments of three vanes each. The shrouds support the segments. The turbine discharge case holds the internal diffuser and the engine rear supports.

The diffuser was designed in order to recover dynamic pressure efficiently while satisfying the overall dimensional requirements.

6. AUXILIARIES

Some of the MS5002E package main features are similar to those of mature gas turbines.

For the main systems, the standard configuration has been evaluated and selected by means of a Quality Function Deployment analysis, considering both Customer needs and internal requirements (technical, manufacturing and commercial).

As with the existing MS5002D, two separate structural-steel frames make the base that supports the gas turbine, one for the engine and one for the auxiliaries. The gas turbine is mounted on the baseplate by means of two forward supports, flexible in the axial direction, and two rear support legs.

The engine baseplate (fig. 4) has approximately the same footprint as the MS5002D and contains both inlet and exhaust plenums. These plenums, with small modifications, are suitable for both vertical and lateral inlet / discharge.

The auxiliary baseplate contains the lube oil system and reservoir, hydraulic oil system, starting system with rotor turning device and fuel gas skid. This modular design will also allow different installation configurations, in order to optimize the plant layout according to Customer requirements.

The MS5002E will use the SPEEDTRONIC Mark VI control system, currently used on all GE gas turbine models.

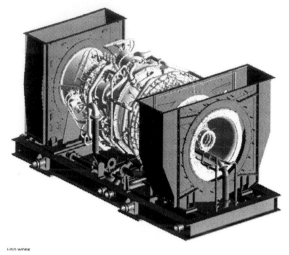

Fig. 4 – Engine baseplate

7. MAINTAINABILITY

Maintainability was carefully considered through a step-by-step analysis of:
- Planned maintenance
- Handling and lifting means for gas turbine components and main auxiliaries
- Special maintenance tools
- Boroscope ports location for easy inspections
- Enclosure access doors and openings

The horizontally split gas generator casings and the removable enclosure roof allow on-site maintenance.

The power turbine is mounted on a special frame that permits its axial displacement on the baseplate. The module can be either disassembled directly on the base, or can be removed for off-base maintenance. Combustors can be disassembled without removal of the compressor discharge case and bearings 1, 3, and 4 are easily accessible for inspection.

The general layout of the auxiliary baseplate ensures good accessibility to the most critical areas and components. An internal crane is provided for main auxiliaries lifting purposes. Filters and instrumentation racks are located outside of the enclosure, in order to make checks and repairs easier.

The planned maintenance intervals are:
- ○ 12,000 hrs - Combustion system inspection
- ○ 24,000 hrs - Hot gas path inspection
- ○ 36,000 hrs - Combustion system inspection
- ○ 48,000 hrs - Major inspection and overhaul
- ○ 60,000 hrs - Combustion system inspection
- ○ 72,000 hrs - Hot gas path inspection
- ○ 96,000 hrs - Major inspection and overhaul

8. TESTS

8.1. Rotordynamic test

A full-scale engine rotordynamic test (fig. 5) was carried out from September 2001 to February 2002 to validate the lateral and torsional dynamic behavior of the entire engine.

The test stand was constituted by the engine (casings, supports, bearings and rotors), two variable speed electric motors, gears, couplings and support frames. The rotors, were the turbine buckets were not installed, featured turbine discs mock up, in order to simulate blades masses. In this way, the ventilation losses are considerably reduced, allowing the achievement of the nominal speeds with low-size electric motors.

The test, preceded by rotors natural frequencies validation, was completed successfully. As predicted, no critical speeds have been detected inside the HP rotor and LP rotor operating ranges, and API616 criteria and limits were met in terms of damped unbalanced analysis and experimental results. Rotors and casing vibration levels have been found within acceptable value both inside the normal operating range and during transients (start-up, shut down).

Fig. 5: Full scale engine rotordynamic test

The test has been organized, following a DOE approach, through several separated sessions where all the parameters that could affect the rotors behavior (oil temperature, bearings clearance, couplings configuration etc.) were alternatively modified or kept constant. The rotors runs of a given session differed for the amount, location and phase of the applied unbalance and also for the speed range of both rotors.

Proximity probes and accelerometers data acquired during test have been analyzed and collected by a software developed by GE Oil &Gas that featured on time trends, bode plots, waterfall diagrams, orbits, spectrum analysis and rotors deflected shapes.

8.2. Compressor Test

A Compressor test started in March 2002 and it is ongoing. The aim of the test was to evaluate both the mechanical behavior and the aerodynamic performances of the compressor.
In particular, the main test activities to be carried out were:
- Define the axial compressor maps on the whole operating range (from 80% to 110% of corrected speed) and at low speed (from 20% to 75% of corrected speed) up to the surge limit.
- Assess the stator vanes and rotor blades actual stresses and natural frequencies through strain gages data analysis.
- Validate rotor and casings thermal model through metal temperature measurements and clearance meters data acquisition.

The test rig development was based on GE Oil & Gas experience.[3]. It consist of the compressor (inlet plenum, casings and rotor) driven by a LM2500/PGT25, an inlet system with air flow measurement tubes, an inlet throttle valve and a compressor air discharge system constituted by three 12" pipes (flanged to the combustor cans flanges) with one valve each.

Fig. 6   Compressor test rig

In order to achieve a given mapping point, the compressor control system regulates the driver speed, the compressor discharge valves opening, and the variable stator vanes actuator position. In some conditions it is also necessary to throttle the test compressor inlet, to reduce pressure and limit the absorbed power within the capability of the LM2500/PGT25 gas turbine.

The test is proceeding with positive results. No major aero-mechanical issues have been detected so far on compressor blades, performances are met both in terms of mass flow and efficiency, and clearance meters data show a good rotor-stator clearances control both at steady state and during transients.

The data acquisition system is based on a server/client structure with a data transfer network. The different clients connected to the main servers enable to elaborate the acquired data through on line trends, scorecards and performance programs. For dynamic data (strain gages, dynamic pressures) on line multiFFT, virtual spectrum analyzer, and waterfall diagrams are also available. Campbell experimental diagrams can be easily post-processed, through dedicated software, for blades vibratory analysis.

8.3. Prototype test
A complete MS5002E prototype test is planned to validate the overall system (engine and auxiliaries). The prototype will use the same engine compressor, previously tested, as driven load. The test will be organized through different operating phases (pre-start, crank, idle, FSFL, base load & tuning, off design) during which the data relevant to the different test objectives (performances, aeromechanics, rotordynamics, thermo-mechanical behavior, combustion, operability/control etc.) will be collected and analyzed.

9. EXPECTED PERFORMANCE

The expected introductory performance @ ISO condition is the following:

LPT shaft speed	6100 rpm
Output (Shaft)	30 MW
SC Efficiency (shaft)	36.4%
CC Efficiency	51.1%
CC Output	46.1 MW
Exhaust Temperature	523 °C
Exhaust Flow	96 kg/s

10. CONCLUDING REMARKS

The MS5002E, the latest development in the GE Gas Turbine family, has been designed starting from the customer needs. The design leverages the Company experience in the Gas Turbine

Design, by means of the use of the NPI Process, of DFSS and of numerous engineering design reviews.

Reliability and availability were considered as key features, therefore extensive use of proven existing design concepts was made. A moderate firing temperature and a high (for a Heavy-Duty machine) efficiency were obtained thanks to a state of the art aero-mechanical-thermal design procedure.

11. REFERENCES

[1] Benvenuti E, "Design and Test of a new Axial Compressor for the Nuovo Pignone Heavy Duty Gas Turbines", Journal Of Engineering for Gas Turbines and Power, July 1997, Vol. 119

[2] Benvenuti E., Casper R., "Development of High-Speed, High Efficiency Power Turbine for the LM2500+", ASME Turbo Expo 1995 Technical paper # 95-GT-410

[2] Mezzedimi V., Nava P., "CTV.- a new method for mapping a full scale prototype of an axial compressor" ASME Turbo Expo 1996 Technical paper # 96-GT-535

Developments in
Pumps and Drivers

C603/023/2003

Ultra-high pressure seawater injection pumps

B GERMAINE
Sulzer Pumps UK Limited

SYNOPSIS

Sulzer have designed and manufactured injection pumps for many years and have achieved many significant references within the industry. Recent deepwater oil exploration, specifically in the Gulf of Mexico, has demanded that a new breed of ultra high pressure seawater injection pumps be developed to meet the needs of deepwater injection.

Recent contracts have allowed Sulzer to extend the range of existing injection pump designs, which has culminated in the manufacture and full verification testing of the World's highest pressure centrifugal injection pump.

INTRODUCTION

Producing hydrocarbons from ultra deepwater is now a requirement of almost all oil companies around the world, with the Gulf of Mexico attracting a significant amount of attention in recent years. The bp fields alone in the Gulf of Mexico, include many of the largest finds in the region. One of these platforms, Thunder Horse TLP will be sited in 1829 metres (6000 feet) of water whereas, Holstein, Mad Dog & Atlantis will be sited in water with a depth ranging from 1219 meters (4000 ft) to 1981 metres (6500 ft). Together these fields have estimated gross recoverable reserves of at least 2.5 billion barrels of oil. This figure could be significantly more if, as now experts predict, the Thunder Horse field yields nearer three billion barrels, three times the original prediction.

Since oil exploration first moved offshore, more than 50 years ago, the meaning of deepwater has changed. The oil industry defines deep water as water too deep for conventional freestanding steel platforms. Today that depth is roughly 400 metres (1,300 feet).

With the need to maximise production from the Thunder Horse field in particular, seawater must be injected into the reservoir at the outset. While this is standard industry practice, it has never before been executed at the injection rates and pressures needed. Overall required injection rates are 200,000 bpd at 586 bar (8,500 psi) and 300,000 bpd at 448 bar (6,500 psi). This is the highest pressure ever achieved within the industry by a centrifugal seawater injection pump.

Taking existing centrifugal injection pump technology and addressing the important factors such as safety and reliability, Sulzer have now designed and manufactured a number of pumps to meet the needs of ultra deepwater injection. This paper discusses the approach taken during the design and manufacturing phases and outlines the important design parameters.

DESIGN PARAMETERS

It is not the intention within this paper to discuss the hydraulic design principles and selection criteria. However, based on our standard designs and experience, the number of stages the pump should have and the size of the pump are easily dictated. For high generated heads, it is common to require a large number of stages. For high heads and relatively low flows, it is common to select our back to back design in favour of the traditional inline pump arrangement. The back-to-back design, opposed impeller pump was designed a number of years ago due to the market demand for high pressure/low flow pumps, in particular for low density product re-injection service. Another design parameter Sulzer define is the generated head per stage, which we limit to 600 metres. This limit has been derived from impeller erosion tests, where to have marginal erosion taking place with duplex metallurgy, the flow velocities through the impellers should be kept to below 55m/s.

Although shaft sizing is not of major importance when designing a pump for high pressure, for most of the pumps that we have either designed or built, it has been necessary to increase the shaft size to a point where our standard hydraulics must be further verified. Larger shaft sizes result in decreasing the eye area of the impeller, which in turn effects the performance of the pump. This concern has resulted in Sulzer undertaking full scale hydraulic development tests before actual hydraulic components are put into manufacture. This procedure was specifically undertaken for the Thunder Horse machine.

For ultra high pressure injection, Sulzer have two designs available. One is the traditional inline arrangement with balance drum assembly to react the generated residual thrusts, see figure 1, and the other is the back to back design with opposed impellers, see figure 2. Both designs have full cartridge withdrawal facility for ease of maintenance. For high pressure designs, it is not possible to use the Sulzer patented 'Twistlock' feature as this design is generally limited to 450 bar (6,525 psi), see figure 3. The higher the end load reacted on the Twistlock teeth from the cartridge, the larger the teeth need to be. This load also results in the barrel diameter increasing to provide better section modulus properties and therefore limit the bending deflection of the matching teeth machined into the pump casing. For pressures beyond 450 bar, the economics direct the designer to the full bolted discharge cover design. For our ultra high pressure injection pumps, all discharge covers need to be full bolted designs, where metal to metal contact can be assured. It should be noted that back-to-back designs, when in operation, have a maximum pressure to be sealed to atmosphere of only half the full discharge pressure.

The table below summarises the ultra high pressure injection pumps that have been fully designed to date. Both the Holstein and Thunder Horse injection pumps have now been manufactured and tested. Both Mars and Atlantis are currently in Manufacture. Table 1 below provides a summary of the pump design and hydrotest pressures. How these pumps rate against the majority of seawater injection pumps that Sulzer have manufactured, can be seen in figure 4.

Table 1 - Ultra Deepwater Seawater Injection Pumps			
Field	**Speed (rpm)**	**Design Pressure**	**Hydrotest Pressure**
Holstein (back-to back design)	5881	586 bar / 8,500 psi	879 bar / 12,750 psi
Atlantis (back-to back design)	6200	630 bar / 9,140 psi	879 bar / 12,750 psi
Thunder Horse (back-to back design)	5820-5997	736 bar / 10,675 psi	957 bar / 13,880 psi
Mars (Inline design)	5320	597 bar / 8,759 psi	776 bar / 11,255 psi

MATERIALS

Barrel/Discharge Cover Manufacture and Materials of Construction
All our ultra high pressure injection pump barrel casings and discharge covers have been manufactured from forged A182 Grade F53 super duplex (25Cr 7Ni 3.5Mo).

For ultra high pressure designs, the main concern relates to the manufacturability of both the casing and discharge cover due the thicknesses of the ruling sections. These are often large as they are governed by the design pressures and barrel diameter. The thicker the ruling section the more critical the heat treatment process becomes, especially for the super duplex. Limiting ruling sections have been established by various authorities based on conventional quench bath techniques. These are the sections that have been achieved whilst still maintaining the mechanical and corrosion properties of the duplex.

If produced water is not specified, there is an advantage in using ASTM A182 Grade F51, due the fact that it is easier to heat treat and obtain the required material p roperties. H owever, ruling sections increase, in the region of 20%, due to the reduction in the material strength.

Carbon steel clad barrels, inconel lined to protect the wetted surfaces from product corrosion, although is probably the low cost solution, has never been adopted for Sulzer injection pumps. The casing wall thickness required to retain the high design pressures are approximately 60-70% larger than an equivalent casing manufactured from F53. Figures 5 and 6 show the pump casings that have been manufactured for Thunder Horse and the casing currently in manufacture for the Mars injection pump. Both these casings are manufactured from F53 super duplex.

PRESSURE INTEGRITY
The first ultra high pressure pump design was undertaken for the Thunder Horse project. Following a full parameter study and risk assessment, a back-to-back design (6 stage plus 6 stage opposed impeller design) running at 6000 rpm was chosen. Injection pumps of this configuration had been designed before but the biggest single challenge was t he generated head required, which w as far higher than any other c entrifugal seawater injection pump to date.

The primary concern was designing for the pressure containment. It was clear that most of our standard design calculation procedures would need secondary review and so it was decided at an early stage, to use extensive finite element analysis techniques. This was the only method available to provide the depth of verification required. The modelling knowledge gained and the level of understanding for how pump components react to very high pressures, has resulted in the pumps designed for all ultra high pressure fields receiving full FE verification. Figure 7 shows some of the various finite element calculations undertaken for the Thunder Horse injection pump.

The most critical components identified are the pump casing (barrel), discharge cover, stage casings and the suction cover, but only if a fully rated suction design is required. For back-to-back designs, the internal interstage cross-over piece has to be evaluated to determine the radial operating deflections (see section discussing static seals).

All our injection pump casings are designed using standard ASME VIII, division 1 code calculations. Full FE analyses of the Thunder Horse casing was undertaken to verify this procedure. Sealing the barrel case to atmosphere and also, due to the large casing thicknessess needed, manufacturing the barrel casing cannot be ignored.

Suction Covers/Casings can be designed to withstand full discharge pressures but FE guidance is needed to ensure that stresses and radial deflections at sealing positions are maintained within limits. We have seen that during hydrotest loading, local yielding of the material takes place at the suction casing shoulder. Material tests with duplex 5A have been undertaken to obtain the exact plastic/elastic behaviour, which is needed when determining the actual component stresses in operation.

The main concern with the discharge cover is the bolt loads induced by the various loadcases and the contact pressure between the cover and the casing to ensure metal to metal contact and therefore pressure integrity. A review of the balance drum liner radial deflections must also be included in the cover analysis, as this item is mounted into the discharge cover. It has been seen from FE investigations that a reduction of the balance piston clearance can be expected for very high pressures and this must be limited by design.

Stage casings within the pump are subject to differing pressure differentials, with the first stage casing seeing full discharge pressure at the outside and approximately one stage pressure on the inside. Due to the high loading on the outside, a reduction in the running clearances can be expected at the wear ring locations. The method of fitting the stationary wear rings can be changed so the radial deflections can be tolerated. Each stage casing has to be assessed individually and it is not uncommon for stage casings within the pump to have differing radial thickness.

Static Sealing
For pumps operating on 100% duty and standby configuration, valve leakage can cause the full discharge pressure generated by the operating pump to be felt by the standby machines. For this fault case, the suction side of the pump will need to be rated for full design pressure and also be subject to the hydrotest pressure rating. Although Sulzer recommend suction relief valves to be utilised in this situation, this is not always confirmed during the design phase.

When s ealing t hese u ltra h igh p ressures to atmosphere, concern is obviously raised and is very much focused on the radial sealing of the suction casing. The discharge side sealing rating is the same for inline pumps, but the methods of sealing are very different. Here metal to metal contact can be maintained by design and so any possibility of any elastomeric extrusion is eliminated.

For back-to-back designs, radial sealing concerns can be focused on the inter-stage piece. Again the radial gap that is generated by the operating pressures, must be limited by design. Note that for this pump configuration, the discharge cover only sees half of the full discharge pressure in operation, although full hydrotest pressures must be sealed.

Standard O-ring seals with backup rings were deemed un-suitable for the required radial seals and so in looking for a sealing solution for the Thunder Horse project, T-shaped elastomeric sealing elements w ith p recision a nti-extrusion r ings w ere s elected f or d evelopment t esting. Tests, with predicted sealing gaps induced, showed no evidence of seal failure, even at 957 bar (13,881 psi). The selected seals have a rateable duty up to 3000 bar (43,500 psi) and it was found during the hydrotests that the seals could be re-used without compromise. For further security, "tell-tale" leakage connections can be added to the pump casing and double seals applied to the suction casing and discharge cover. This design allows the pump operator to know if the primary seal has failed. Figure 8 shows the "tell-tale" connection that can be added to monitor a discharge cover primary seal.

Low Pressure Suction Design
It is important to quantify the issues that surround the pump suction being designed for the full discharge pressure. Sulzer strongly recommend that the pump suction should be designed for suction pressure magnitudes only, although this recommendation is not always accepted by the end user. The problem stems from the fact that the platform process pipework is not always being designed at the same time as the injection pump, and so a pump designed for full discharge leaves options open. However, the benefits of having a low pressure suction design are as follows;

- The pump case wall thickness at the suction end is minimised, which generally brings the section thickness to within known limits for forged barrels.
- For ultra high pressure injections pumps, there is a massive weight saving. Sulzer have shown that a reduction in the overall barrel weight by as much as 25% can be achieved, just by reducing the design pressure at the suction side to normal suction pressure rating.
- Distance between bearings can be reduced, thus improving rotordynamic behaviour.
- The maximum pressure that mechanical seals can tolerate, as a fault condition is approximately 250 bar, without the need to replace the seal faces.
- The optimum mechanical seal solution can only be provided if a low suction pressure design is chosen.

Figure 9, shows the comparison between the pump case design for both high pressure and low pressure suction designs. Additional cost would be removed if the suction casing, mechanical seal plate and seal bolting could all be designed for normal suction pressure magnitudes.

MECHANICAL SEALS

Mechanical seal selection for ultra high pressure and high speed injection pumps, clearly needs careful evaluation. Other factors such as suspended solids contained within the pumped medium also need to be taken into account. The industry standard for applications were sand is present within the pumped media, and specifically produced water applications, are pressurised dual seals with a back-to-back arrangement. However, for the Gulf of Mexico platforms, where reverse pressure requirements at one time were stated at 10,000 psi but finally were set at around 3,600 psi, the John Crane tandem design was considered to be the preferred dual seal arrangement.

Double mechanical seals operate at high pressures and speeds and therefore, there is a quantifiable loss of the barrier fluid to both the product and to the environment. To minimise this loss, it is essential to optimise the seal face design and to ensure stable operating conditions in terms of both pressure and temperature.

Based on this requirement, it is very important not to over rate the suction design pressure. If the suction end of the pump can be rated to the normal suction operating pressures, then the mechanical seals can be optimised for these operating pressures to minimise barrier fluid leakage. For safety considerations, the seal plate and bolting can always be designed relatively easily for the full reverse/discharge pressure ratings.

ROTORDYNAMICS

It is our standard design practise to undertake fully damped lateral vibration analyses for all injection pump designs. There is no change to the procedure when the rating of the pump is for ultra high pressure. Depending upon the duty requirements, it may be possible to select either an inline or back-to-back design configuration (see figures 1,2), but the back-to-back design should be selected for abrasive wear applications as they are less sensitive to pump wear due mainly to benefits provided by the centre bushing. The pressure breakdown across this bush is half of the full discharge and so relatively high stiffness and damping magnitudes are generated. From a rotordynamic point of view this centre bushing is a very effective hydrodynamic bearing.

All impellers, shaft sleeves, balance piston, coupling hub and thrust collar are shrunk onto the pump shaft. This improves rotordynamic performance significantly by keeping shaft vibration levels to a minimum. Shrink fitted components also promote repeatability of rotor balance for future builds.

The criterion employed is full compliance to API 610 8th Edition, Appendix I. Therefore, Eigenfrequencies and modal damping levels are computed for set speed intervals covering the desired speed range, for both new (clearances as design) and worn (2x design) clearances, see figure 10.

One of the most important parameters to examine is the pump coupling weight. Sulzer have a specific procedure that allows us to examine what we call "Coupling Sensitivity". Sulzer perform this check calculation because often Eigenfrequency calculations do not provide sufficient information on whether a particular design and coupling overhung mass are acceptable. Any overhang dominated mode may be far removed from shaft speed frequency, but still unacceptable vibration levels may show up in operation.

The rotordynamic design and corresponding stability margins are very important when considering pump reliability. Understanding the change to the rotordynamic behaviour as the pump wears has to be analysed accurately. For the Thunder Horse project, bp undertook to verify the rotordynamic design calculations and specified unbalance response tests to be undertaken during the pump prototype testing for both the new and worn clearance conditions.

Bode plots taken at the end of the official string test for the prototype machine are shown in figure 11. The 1x filtered measurements shown, have been compensated for slow roll. They show that the machine operating at full speed has very low levels of shaft vibration and also there are no rotordynamic resonances from full to zero speed.

ABRASIVE WEAR

All seawater injections pumps are expected to pump a percentage of suspended solids, normally quartz sand, during their productive life. For produced water injection, where significant quantities of sand can be present, the pump designer must protect specifically the close running clearances against rapid wear.

Experiences in Alaska in the early nineties [1] and in more recent years experience from the North Sea, have given Sulzer the opportunities to include abrasion resistant materials and component designs. The most vulnerable components of any high pressure injection pumps are the close running clearances, because here, the velocities are high and for inline arrangements, the critical item is the balance piston.

It is important for long term, reliable operation, that the balance piston and liner are provided in materials that are sand resistant. Any increase in the clearance away from original design, will result in higher leakage, a reduction in pump efficiency, a modification to the residual thrust and possibly an over-pressurisation of the balance piston return chamber, although in most cases this can be eliminated by careful design. Back-to-back designs, because they do not have a single full discharge break-down piston arrangement, generate much lower fluid velocities across the long bushings. Because of this fact, a back-to-back pump design should be selected whenever possible for abrasive service injection applications.

For balance pistons/liners and throttle/centre bush arrangements, severe abrasion resistant designs exist. For sleeve components, a sintered tungsten carbide sleeve is bonded to a molybdenum carrier which is used to mount the whole assembly on to the shaft. It is normal for the balance piston in our HPcp injection pumps to have a heavy shrink fit, which is used to seal the bore to pressure/leakage and to provide the required drive. When using tungsten carbide, which is basically weak in tension, this method of assembly is not possible. In all our sleeve assembly designs, full finite element analysis for all envisaged loadcases needs to be undertaken to ensure that the tensile loads applied to the carbide sleeves can be accommodated. Tungsten carbide liner designs are much simpler to arrange, as the carbides all have extremely high compressive strengths. Separate liner sections are stacked together, so features such as the Sulzer deep radial grooves and circumferential swirl brake features can be pre-machined into each of the liner sections, see figure 12.

In the last few years, third generation HVOF thermal spray coatings have been developed that allow the pump design engineer to provide a level of protection without the need for complex carbide assemblies. Clearly solid carbide designs provide a high level of erosion resistance but, are generally used when all other measures have failed. However, their cost, weight and the fact they are difficult to manufacture, means that other methods of protection need to be available.

For the produced water injection pumps supplied to Maersk Olie & Gas, Sulzer Pumps along with Sulzer Innotec and Sulzer Metco undertook a detailed development programme that assessed commercially available HVOF coatings against the latest Metco powders and third generation deposition methods, where deposition velocities are much higher. Two body and complex abrasion testing in a specially designed "Heli" rig where undertaken, which allowed all materials tested to be ranked. The "Heli" test facility, also allowed each of the materials tested to be exposed to impingement erosion at different angles of impact. It can be seen in figure 13, that all tungsten carbide materials (Homogenous or HVOF applied) have lower wear rates when impact angles are high. This is due to the fact that these materials are all brittle in nature and therefore suffer higher erosion rates when impingement angles are high. Duplex and weld overlays are ductile in comparison and the effect of this can also be seen in the diagram.

The results of the development programme showed that solid sintered tungsten carbides are, by an order magnitude, more abrasion resistant than even the best HVOF coatings. It also showed that the latest Metco SUME®*PUMP* HVOF coatings provide better protection than all other HVOF coatings tested. These coatings have now been applied to many of our produced water injection pumps operating in the field with excellent results and this material is being selected for many ultra high pressure injection pumps in order to provide protection and therefore enhanced pump life. Sulzer Metco are capable of providing SUME®*PUMP* coatings that can be deposited to 1mm thickness and thus meet the requirements of API 610 8th Edition. Further information on abrasion protection and mechanisms can be found in [3].

NOZZLE CONNECTIONS
For all our ultra high pressure injection pumps, Techlok clamp connectors are preferred over conventional flanges. For the Thunder Horse pump, standard ASME B16.5 type flanges could not be used due to limiting pressure ratings. The Techlok connector is a compact mechanical joint consisting of hubs, a seal ring, two half clamps and bolting. Unlike conventional flange designs, the clamp connector does not rely on the bolting to maintain seal integrity during service. Sulzer have a lot of experience with these type of connectors and are preferred even for the lower pressure injection pump designs.

The main benefit to the pump manufacturer is the smaller and lighter pump nozzles. For the operators, close piping and hook arrangements and general ease of maintenance are the main benefits. Techlok clamp connectors can deal with both thermal and mechanical shocks and pressures even up to 2069 bar (30,000 psi) are possible. The joint can also be reused after assembly and will seal statically once the clamps and lateral bolts have been tightened.

TESTING

Hydrotest
Due to the very high hydrotest pressures required for the Thunder Horse prototype pump (957 bar), local visual examination and leak detection was not possible. After a full risk assessment, the pump hydrotest assembly was positioned vertically in an enclosed concrete pit. Four strategically mounted video cameras, aided by the use of a special fluorescent dye, were used for visual surveillance. Pressure transmitters reading any leakage from the "tell-tale" ports were used, including digital DTI's applied to the casing outside diameter and the suction casing to monitor expansion and elongation respectively.

The balance return pipework was also pressure tested as part of the combined verification. Once full hydrotest pressure is achieved, it is normal to hold the pressure for a set period. For Thunder Horse, full hydrotest pressure was held successfully for more than thirty minutes. The full pressure test arrangement can be seen in figure 13. Measured casing expansions and the suction casing elastic elongation compared very well with the FE investigations carried out to verify the pump design.

Prototype Testing
Due to the fact that the Thunder Horse injection pump is the World's highest pressure centrifugal injection pump ever manufactured, extensive development tests of the bare shaft pump were undertaken. The pump included many significant advances in design, most being introduced as a direct consequence of the high pressure. As well as hydraulic performance tests, the client stipulated unbalance response shop tests at full speed and with the pump having design, then, worn (2x design) clearances. This was managed by fitting special sacrificial stationary wear parts.

Full mechanical testing with applied weights to the pump coupling did not promote rotor instability, even when the pump was fitted with worn clearance bushings. The results from the tests fully endorsed the rotordynamic lateral analyses. The bare shaft pump being slave tested and the pump being string tested as a complete package can be seen in figures 14 and 15.

BIBLIOGRAPHY

[1]   Solutions to Abrasive Wear Related Rotordynamic Instability Problems
       on Prudhoe Bay Injection Pumps
       Robert A. Valantas and Ulrich Bolleter
       5th International Pump Users Symposium

[2]   Innovative Upgrades to Address Cavitation, Erosion and Rotordynamic Issues
       in a High Speed Diffuser Pump
       Brien Jones

[3]   Extension of Pump Life in Abrasive Services
       Paul Meuter and Richard K. Schmid
       Pump Users International Forum 2000

[4]   Deep Thinking
       Terry Knott
       Frontiers September 2001

C603/016/2003

Ultra-high pressure water injection pumping design

N MUNTZ and **G STEAD**
Weir Pumps Limited, UK

SYNOPSIS

With an ever increasing drive to find oil the requirements for high pressure reservoir maintenance has been steadily rising with up to 600 bar required for continuous well fracturing and injection.

A single pump solution for this application has been developed with the key objectives of achieving: -

- A high level of safety requiring centrifugal pump technology to be developed to a standard exceeding that currently available in the market place.
- A high through life efficiency.
- An expected service life of 25 years.
- A pump design suitable for continuous operation at pressures up to 600 bar, compatible with both seawater and produced water injection.

With its long history of designing engineering solutions to fluid pumping problems, a Weir pump design was developed from solid proven fundamentals for the required application rather than adopting a stretch approach to an existing design. Basic hydraulic and mechanical design fundamentals were therefore established for the envisaged operating range and duty requirements.

This paper discusses the pump selected together with the process and methodology followed to arrive at the final solution.

1   INTRODUCTION

Recent years have seen a decline in the natural mineral resources available in existing oil wells. This has prompted a move to drill deeper wells, in deeper water at increasingly remote and extreme locations. These factors have seen the requirements for reservoir pressure maintenance rise to levels in excess of 450 bar.

Current methodologies adopted for achieving these pressures involve series pumping. This, although successful, has a number of inherent difficulties:- topside equipment requirements and capital cost increases, it is necessary to offload the pressure applied to the mechanical seals on the high pressure units to an acceptable level and overall efficiency levels are reduced together with an increased maintenance load and resultant decrease in reliability.

This paper details the process undertaken in producing a robust design solution to enable the replacement of current series pumping methods with a single ultra high pressure unit.

The paper will summarises the methodology followed in arriving at a final design solution, the key features to be addressed and a description of the final solution arrived at.

2   APPROACH

2.1   Design process

Traditionally, new pumping solutions are arrived at through an evolutionary process from existing designs. Whilst this approach provides a useful route towards steadily increased operational boundaries, there invariably comes a point where new philosophies need to be pursued. Ultra high-pressure operation can be approached in this manner through series pumping or a 'stretch' design, however an alternative solution was sought in this case utilising existing technology in a novel arrangement.

This approach would allow for a design optimised to the new duty with future expansion available in the event of even higher pressures and/or flows being required in the field.

Basic hydraulic and mechanical design fundamentals were established for the required operating range and developed to suit high-pressure duty requirements. An iterative process was then followed during which the design was refined to generate the best and most flexible solution.

As the design project progressed key characteristics were identified which allowed comparison of each proposed feature, leading through from intermediate design models to the final solution.

2.2   Basic Design

To achieve typical injection flow rates at the pressures required, a single multistage pump operating at high speed is required. Although various design permutations are available to the designer, common practice dictates that pumps operating at speed's equalling approximately 6000 rpm require a unit with at least nine stages to generate the required necessary head without compromising efficiency. Higher speed options with fewer stages can also be utilised but yield little advantage. The design concept developed was therefore based on a nine-stage unit with 340 mm impellers, operating at design speeds of 1.5 times the operating speed.

2.3 Comparison of design solutions

As the design process progressed a number of alternative solutions to the original design specification was generated. Weighting methods to eliminate non-preferred designs were incorporated into the design process. Identification of key pumping features early in the product development aided in this process and also formed the basis of a risk analysis leading to the final solution.

2.3.1 Identification of key features

By analysing many years of operational history and site experience, a key list of features affecting previous site operability was drawn up. Through life maintenance records from the North Sea have provided particularly valuable information with respect to harsh operating environments.

1 Mechanical Seal Selection	The principle cause of down time in an operational environment is failure of the mechanical seal and/or seal system. To ensure maximum reliability, the seal environment must be carefully controlled and selected.
2 Bearing Selection	High-speed application requires stable bearing regimes and suitably sized thrust capability to increase the ability of the pumpset to tolerate minor oil supply variations and improve reliability.
3 Rotor Dynamics	High-speed multistage application requires stable wear conditions to maximise through life availability. It has long been recognised that minimum shaft deflections and vibration characteristics are required to ensure an ideal environment for pump components. This is best achieved through a stiff shaft design with stable rotor dynamics.
4 Wear Rates	Sand content of produced water is unknown and wear rate condition is critical to overhaul periods and efficiency - to allow for operating in Produced Water. The wear rates must be considered.
5 Pressure Integrity of the Design	The high pressure rating exceeds known experience in injection water applications
6 Ability to Manufacture	High pressure rating requires heavy construction of main pressure containment, exceeding normal experience in duplex water injection applications. Additionally, the number of stages required to generate the required head efficiently must be matched with a proven manufacturing route to ensure full repeatability of build and resultant rotor balance.
7 Through Life Cost	A major benefit to end user with respect to both OPEX and environmental considerations

2.3.2 Risk analysis of the key features

The key features of the pump design discussed above were plotted on a risk assessment chart as shown in figure 1. The assessment of each design feature was based upon its impact on reliability a nd p latform revenue and the degree of difficulty assumed in risk management. Areas of key concern were identified, associated with those features having a high score against both axes on the Risk Analysis Chart and are -

- 3 - Rotor Dynamics
- 4 - Wear Rates
- 6 - Ability to Manufacture

Mechanical seals (1), bearing selection (2) and through life costing (7) have a large impact on reliability, however, prior experience in these areas can be readily utilised and managed. Pressure integrity (5) at these levels, whilst novel within the field of centrifugal pump manufacture, can be managed through diligent use of finite element analysis and experience in other environments.

The design options were assessed against the key, critical design features to arrive at the optimum solution.

3   DESIGN SOLUTIONS

The Weir range of centrifugal multistage barrel pumps featuring full cartridge withdrawal facilities was taken as a general basis for the new pump design.

Specific elements of this design were then analysed with variations generated to meet the performance requirements of ultra high-pressure applications.

These elements were used to develop numerous alternative pump designs, which were assessed against the areas of key concern to arrive at the final optimum solution.

3.1   Key design elements
- Pressure containment and Suction Approach
- Hydraulic element arrangement
- Product lubricated bearings

3.1.1   Pressure containment and suction approach

Two pressure containment options are commonly used on barrel pumps operating at high pressure these being a shear ring/retaining ring and a bolted end cover arrangement.

The shear ring/retaining ring arrangement distributes the axial load throughout the whole of the end cover barrel perimeter via a 360-degree shear ring and has been successfully used on many cold-water applications. The inner pump assembly is held in a loaded position by the use of an axial compensator, in the form of a disc spring, against the non-drive end cover. This "cartridge lock" type system eliminates the need for heavy casing studs, additional sealing flanges and hydraulic tensioning systems and contributes to significant reduction in cartridge replacement down time.

In the case of the bolted end cover arrangement axial loading is distributed through the equi-spaced discharge studs. These studs are hydraulically tensioned to generate pre-load and prevent any distortion effects arising from pressure and/or temperature.

In addition, the bolted arrangement allows face 'O' rings to be fitted to the main pressure containment joint. This provides added security against leakage at the bore' O' ring caused by extrusion effects at extreme pressures.

The bolted end cover arrangement was therefore proposed as the method of pressure containment and was used in the development of all pump options.

Whilst an acceptable method of securing the endcovers can be readily established, local distortions and high stress distributions will be created where penetrations are made through the barrel.

To improve pressure and static sealing integrity, it is therefore desirable to position penetrations as far removed as possible from the endcovers, particularly with respect to the larger suction opening.

Three methods for suction branch locations were analysed, these being a conventional arrangement (located at the pump drive end), the suction guide positioned over the impellers, and a central suction guide.

Conventional pumpset arrangements have the suction guide approach contained within the end cover, with a smooth sweep directing the flow of injection water directly into the impeller eye. This provides good hydraulic flow into the impeller eye and is acceptable for conventional pump heads and speeds.

A modification of this arrangement allows for the suction guide approach being positioned over the rotating element cartridge with the flow directed over the impeller shrouds in a smooth arc into the impeller eye. This allows the pressure breakdown bush and seal area to be recessed into the suction cover, thus reducing bearing centres and improving rotor dynamics. Additionally the barrel penetration for the suction passage is then located away from the barrel end reducing distortion effects at the critical pressure boundary sealing location.

To allow further improvement the suction guide inlet can be brought centrally into the barrel via a front-to-front opposed impeller design. This builds on the advantages of positioning the suction guide over the impellers and ensures that the critical pressure boundary seals are located in a fully axi-symmetric area of the barrel.

This completely removes any unwanted distortion effects and ensures that pressure integrity is optimised. This is the preferred arrangement and is utilised where possible in the concept designs. By routing the discharge flow back towards the barrel centre, the sealing of both end covers can be made fully axi-symmetrical, providing greatly increased performance with respect to the main static pressure containment.

3.1.2 Bearing stiffness and wear rate considerations.
Due to the high operating speeds of the pumpset, the rotor-dynamics cannot be compromised. To cope with the rigorous pumping demands whilst promoting safety and reliability a stiff

bearing philosophy is recommended whereby the rotor has high damping coefficient, minimal deflections and good wear tolerances. The most effective way of ensuring that these factors are obtained is to promote a good flow of product fluid through the bearing by maintaining good pressure differentials at all times.

The unknown quality of sea water/produced water, with possible entrained solids also requires that any potential effect on wear needs to be evaluated. Wear is caused by either two or three body mechanisms as illustrated in figure 2.

Two body wear is caused by impingement of particles within the pumped fluid onto the bearing surface. When this mechanism takes place the critical factors are velocity, angle of impingement and hardness of the material.

Three body wear is caused by the "rolling" of particles between the two bearing surfaces, and is typically found in sand and silt laden fluids.

Under this type of erosion even hard materials will experience wear regardless of the velocity of the fluid as hard particles are ground into the bearing surface. Key factors associated with this form of hard particle wear are therefore particle size and residual clearance.

Under normal circumstances three body wear will occur during early running of the equipment, this will gradually diminish as residual clearances are worn out to greater than particle size. Eventually three body wear will cease and two body wear will become the critical mechanism. See figure 3. Preventing both mechanisms ensures adequate through life resistance.

Bearing fluid film stiffness is maximised by ensuring a high-pressure differential across the bearing ends; controlling the rotor deflections and eliminating three-body wear. Hard materials are utilised at all bearing surfaces eliminating two body wear.

3.1.3   Hydraulic element arrangement
Three possible arrangements for the impellers were assessed for the proposed design, these being an in-line arrangement, a back-to-back opposed arrangement, and a front-to-front arrangement. These arrangements can be seen by referring to figure 4.

In all arrangements the impellers, balance piston and thrust collar are shrunk onto the shaft and keys are fitted to transmit the torque as per conventional equipment. The thrust generated by each impeller is also partially opposed by the balance piston leaving a small residual thrust to be carried by the thrust bearing.

A number of key design elements can only be utilised in specific hydraulic element arrangements and these constraints are discussed below: -

In a conventional in-line arrangement all impellers are fitted to the shaft in the same direction with a product lubricated bearing fitted between the second and third stage impellers to improve rotor rigidity. This type of arrangement is standard for radially split barrel pumps because of its ease of build, simple construction and reliable robust proven rotor dynamics at conventional duties.

   C603/016/2003 © With Author 2003

With this arrangement it is not possible to fit the preferred options of a centrally arranged suction guide or a high-pressure product lubricated bearing and in effect this is an example of a 'stretch' design where future proofing is not possible.

As an alternative, a back-to-back arrangement where the impellers are positioned in two banks with the suction of the impellers oriented away from the centre bush of the pump can be used. This arrangement balances the axial thrust of the two impeller sets thus reducing the overall load on the thrust bearing and is commonly utilised in low pressure axially split casing pumps where access is available to the impeller hub for assembly/disassembly. With a barrel construction the internal cartridge is arranged with radially split ring sections. With this arrangement, the impellers are assembled and remoed from the eye direction and this, combined with the use of a radially split internal cartridge arrangement prevents access to the rear of the impeller. To ensure accurate location to the shaft a shrink fit is utilised in this area and accurate and repeatable assembly/disassembly therefore becomes extremely arduous due to this lack of access. Repeatable rotor assembly and balance cannot then be achieved.

With this arrangement it is not possible to fit the preferred options of a centrally arranged suction guide or a high-pressure product lubricated bearing at the drive end location and inherent limitations are thereby imposed on the complete pump design.

To overcome these difficulties, the back-to-back arrangement can be further modified by mounting the two banks of impellers conventionally with the eye orientation towards the middle of the pump. This allows the impellers to be assembled and removed from the hub direction giving full access and allowing a conventional build approach to be adopted, significantly increasing reliability of original build and ease of maintenance.

With this arrangement all product lubricated bearings have a natural high pressure differential across them and the rotor is therefore inherently stable with all natural frequencies critically damped and removed from running speed. This ensures that through life wear is minimised and through life cost is therefore reduced.

A centrally arranged suction guide is also utilised with this arrangement, which when combined with the bolted end cover arrangement provides maximum pressure boundary integrity for the extreme pressures of ultra high-pressure water injection.

This combination of features therefore provides a complete solution to all the required design features and can further be conventionally modified for improved wear resistance of sand is known to be present in the pumped fluid.

3.2 Optimisation of Impeller Split
Throughout the design process, all opposed configurations were considered with the impellers arranged equally each side of the central bush. This naturally results in a balanced thrust design, but does not readily allow pump rotor characteristics to be tuned. Varying the number of impeller stages each side of the centre will result in an alternative set of rotor designs in which the thrust can be equalised by differential sleeve diameters.

Alteration of the impeller bank configuration enables an increased level of control over the efficiency and rotor dynamics of the pump. This enables individual tailoring of the pump to suit specific needs depending on possible future customer requirements.

The final arrangement selected then depends on the precise duty requirement and is primarily selected based on the characteristics of the bushes used to stiffen the pump shaft. These bushes operate in the pumped fluid and provide greatly increased rigidity to the shaft system dependant on the pressure driving the fluid across the bushes, where an increased differential pressure results in an increased stiffness and hence improved rotor stability.

Analysis of the rotor with the bushing at a number of different impeller positions must therefore be carried out to optimise the relationship between rotor rigidity and pump efficiency.

For example arranging the impellers with an unequal number of stages each side of the centre can improve efficiency. Splitting the impellers evenly increases rotor dynamic performance by stiffening the shaft resulting in improved reliability, safety and through life performance whilst reducing vibration characteristics.

Having optimised the position of the bushing, the sleeve diameters used are then selected to ensure minimum thrust loading on the bearing.

For the purposes of analysis and discussion, a split of 6-3 impellers was used (see figure 4), with the bias tending towards the drive end. This means that the impellers are located between three hydrostatic bearing bushes, with three impellers between the non-drive-end and centre bearing bushes and six between the centre and drive-end bearing bushes. This configuration gives an efficiency slightly higher than that obtained when splitting the impeller banks into a reference 4-5 configuration due to the reduced pressure breakdown across the none drive end and centre bearing bushes: three stage pressure only being broken down as against four stage for the reference case.

3.3 Rotor dynamics

Rotor dynamics of any pump form a critical factor in the performance and reliability of any pump. During the design of the Ultra High-Pressure pump the rotor dynamics were continually assessed to ascertain the impact of any changes to the configuration of the impeller banks. Due to the nature of the opposed impeller design it was assumed that stiffness of the pump shaft had been increased "naturally" due to the additional hydraulic support afforded by the inclusion of a third hydrostatic bearing. This effectively reduced the applied mass at the centre of the shaft, where catenary displacement is highest, by dissipating the load. The opposed impeller configuration also reduces the effects of axial thrust by applying the load against itself. Other factors assumed to aid in stiffening the shaft were the hydraulic forces produced by the impellers and diffusers, which can be said to "lock" the shaft into position.

The Rotor Dynamic analyses performed by Weir included a look at both the lateral and torsional effects induced by the rotor. Lateral analysis considered displacements created by the pump-rotating element at up to 150% of nominal speed. Campbell and damping diagrams to API 610 8th edition were produced to ensure that the pump would perform at the necessary international standards. Consideration was also given to worn clearances at two times normal. Lateral analysis gives a clear indication of the mode shapes generated by the pump under running conditions indicating areas of the pump which may impinge on any design clearances which cause the pump to seize or reduce the through life performance.

Torsional analyses allow stresses in the shaft to be checked to API tolerances, again allowing an indication, at the design stage of the project, of the induced stresses. This allows modification of the design to achieve optimum performance early in the design process.

As predicted the optimum rotor dynamic configuration would be to have the impellers split uniformly a cross t he c entral h ydrostatic b earing, i .e. the 4 -5 s plit. F or d iscussion purposes results from the 6-3 split configuration at normal clearances are included in figures 5,6 and 7.

The results are plotted to a speed of 9000rpm which is 50% above operating speed. The highest response at the operating speed is about 1.2 Mils (Pk-Pk) for design clearances. Response at the bearings is <0.4 Mils for the design clearances. Given the G90 level of forcing applied, these responses are excellent.

The stability analysis plots real (damping), and imaginary (natural frequency) parts of the Eigen-values with running speed. These show the critical speeds, their level of severity, and at what speed the stability threshold is reached, i.e. the rotor becomes dynamically unstable. Under these definitions it is the case that a highly damped critical speed can exist which is benign. Reference to figure 7 shows, the running speed at which the dashed line is crossed is a critical speed, but account must also be taken of the level of damping. For example, API 610 8th edition allows a critical speed with a damping level greater than 15%.

In this analysis, more than 20 modes are calculated, some of which are negative precession modes that do not excite under normal conditions. These and other highly damped modes are not plotted for brevity. The forward precession modes up to the third bending mode and below 12000cpm are plotted and described in the legend. The key issues are any critical speeds with low levels of damping, of which there are none, and the level of damping in the modes around ½ running speed. These are the modes, which tend to control the stability threshold, for which damping levels are very high.

The Opposed design is excellent in terms of both response and stability. Response levels due to impeller unbalance levels are limited. This largely stems from the internal hydrostatic bearing supports, effectively giving a substantially shorter span between supports. The pump response is seen to be insensitive to wear, so given the high values of unbalance used in the analysis, the resulting low values of shaft response, and the limited wear of the tungsten carbide wearing surfaces it is clear that the design is rotor-dynamically robust.

4   FINAL DESIGN SOLUTION

The water injection pumps are high-speed units of barrel casing/cartridge design with face to face opposed impellers.

After r emoval o f t he p ump c oupling s pacer and a ncillary p ipe w ork, t he cartridge may be removed from the barrel casing without the need to break the suction and discharge branch connections.

The cartridge consists of a Drive End (DE) and Non-Drive End (NDE) journal bearing assembly, an intermediate cross over housing and two conventional impeller/diffuser/ring section stacks.

Mechanical seals, which prevent fluid leaking between the rotating element and stationary casing, are housed in the end covers.

The DE and NDE journal bearings are from a standard range of bearing housings as used on all of our high-speed units. These are pressure fed with cooled oil from a separate lube oil system.

5   CONCLUSION

The Weir Opposed front-to-front impeller pump design is a robust design solution to ultra High Pressure Water Injection pumping, which utilises existing components in a revised layout to produce a dynamically stiff rotor with additional key features including:–

- Face 'O' rings on key pressure containment joints
- Centrally located suction and discharge branches for minimum distortion
- High pressure product lubricated bearings
- Potential for hard material wear surfaces
- Face-to-face opposed impeller design for maintainability
- Expected mean time between overhaul of 5 years

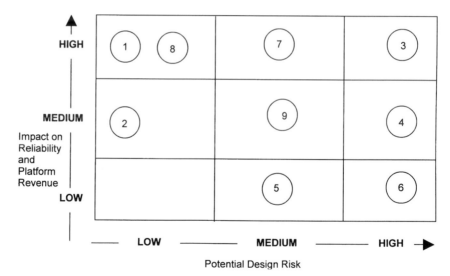

Figure 1 – Risk Assessment Chart

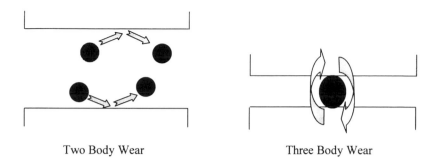

Two Body Wear   Three Body Wear

Figure 2 – Wear Body Mechanisms

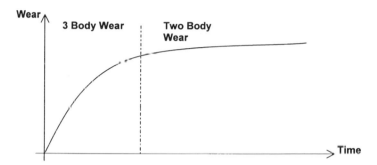

Figure 3 – A Graph Demonstrating the Effects of Wear Over Time

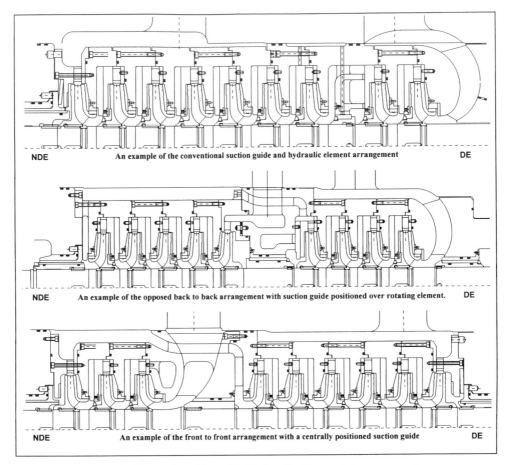

Figure 4 – Examples of Possible Hydraulic Element Arrangements

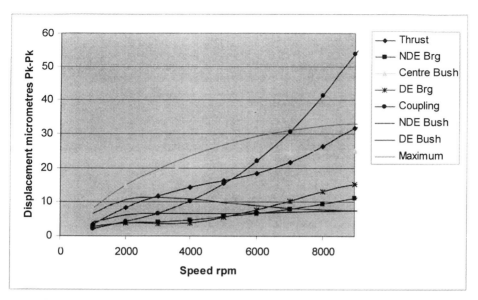

Figure 5 – Opposed Pump Displacements Against Speed at New Clearances

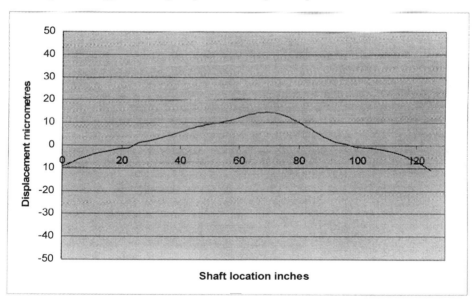

Figure 6 – Opposed Pump Shaft Displacement at New Clearances

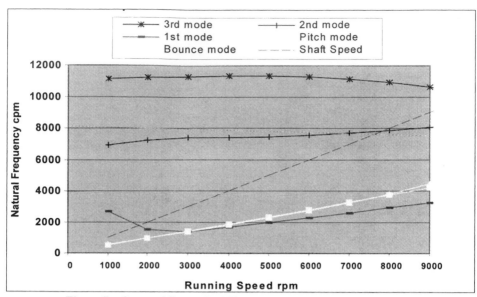

Figure 7 – Opposed Pump Stability Analysis at New Clearances

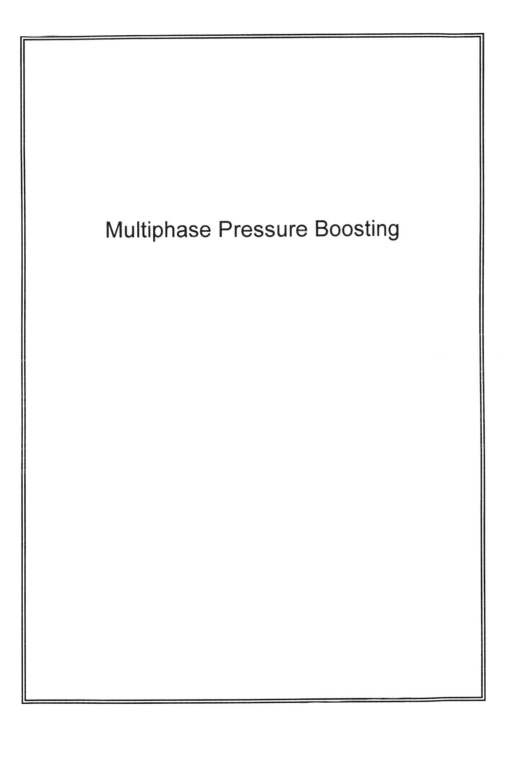

Multiphase Pressure Boosting

C603/014/2003

Field operation and performance of a downhole hydraulic submersible multiphase pump

W G HARDEN
Weir Pumps Limited, Glasgow, UK

SYNOPSIS

This paper describes the design, field deployment and operation of a unique hydraulic turbine driven submersible downhole pumpset for multiphase boosting of hydrocarbon wells. The pumpset is an evolutionary development of the hydraulic turbine driven pump concept, which was first applied to well boost duties in the mid 1980s. The successful deployment of the new design in a subsea field has shown that continuous, reliable operation can be achieved in a downhole multiphase produced fluid environment.

1. INTRODUCTION

In the production of hydrocarbons, wells often require boosting to achieve economic recovery quantities and rates. Downhole pumps are therefore used to artificially boost the producing pressures to drive the fluids to the surface and to provide sufficient pressure for onwards transmission to production facilities. Co-production of gas with the well liquids is a fairly common occurrence, which can pose problems for conventional electric motor driven centrifugal pumps. However the requirement to operate pumps in gassy applications will increase as we continue to produce wells at lower suction pressures at or below the oil bubble point in order to maximise oil recovery.

The problem of pumping liquids containing significant quantities of free gas has long been recognised as one of the limiting factors in downhole pump availability and reliability, due to the performance constraints, additional equipment requirements and variable loading it imposes on pumps and their associated electric motor drive systems. Traditionally, in-line rotary gas separators have been used in well pumps to try and maintain pumping when significant gas is encountered. However, in recognition of the potential for improved performance, Weir and ChevronTexaco collaborated in the development of a variant of the Weir hydraulic turbine driven submersible downhole pump, (HSP), incorporating an improved stage design for handling high gas fraction fluids, (1). This programme concluded with the development of the world's first high volume rotordynamic multiphase pump for well applications.

Following the completion of a successful offshore live well field trial of the new HSP, (2), the decision was made to go for full field deployment of the technology in the development of ChevronTexaco Captain Area B. The Captain Area A development, which came onstream in

March 1997, employed conventional electrical submersible pumps, (ESPs), deployed in wells accessed from the main platform. However, development of Area B - the eastern part of the field - presented a further challenge with the presence of an extensive free gas gap which would prove problematic for ESPs or gas lift, the most common offshore production lift methods.

A significant driver for deployment of the new technology was the desire to develop the field extension on the basis of subsea facilities, with no ready access for equipment replacement in the event of premature failure. It was considered that the hydraulically powered HSP system would maximise availability and reliability when compared to electric ESPs, with a corresponding reduction in well workover frequency and costs.

2. HSP PRINCIPLE OF OPERATION AND ARRANGEMENT

The hydraulic turbine drive concept utilises high pressure liquid as a drive medium to power a multistage axial flow turbine mounted on a common shaft with a multistage pump. Power fluid, either water or oil, is supplied from the surface using a charge pump and most commonly mixes with the pump discharge fluids at turbine exhaust, the two fluid streams commingling and flowing into the production system for subsequent separation and processing, **Figure 1**.

The HSP is run on production tubing and located within the well casing at a depth determined by suction pressure requirements. Power water is supplied down the well tubing / casing annulus and enters into the turbine through a junk basket debris screen. In the Captain application the well completion system incorporates a bypass for well logging, which can also be used as a conduit for un-pumped production by gas drive in wells with very high gas content, **Figure 2**. A novel arrangement of valving was employed to isolate the HSP when producing the well naturally, in order to minimise "windmilling" and deposition of solids within the pump internals.

The successful long term performance of HSPs is down to three critical success factors:

- Scale control   Power water and produced fluids compatibility must be maintained to avoid scale precipitation and build-up

- Power water solids control   Solids in the power water system to be controlled to within standard injection water quality specification of 80 microns and 2 ppm by volume

- Produced fluid solids control   Produced solids to be less than 100 microns maximum particle size and 10 ppm by volume, (ChevronTexaco Captain sand screen design basis is 2 ppm average and 200 ppm upset)

3. DESIGN FEATURES

A new design of pump stage was the key factor in the development of the multiphase HSP, facilitating the successful pumping of liquids containing high volumes of gas. Conventional mixed and radial flow pump impeller stages have a tendency to centrifuge the gas and liquid fractions, with the result that the pump effectively ceases to draw fluid through the stages and generate head. This condition is known as "gas locking" and when it occurs it usually requires the operator to shut down the pumpset to allow the gas to pass through the pumpset and the pump to re-prime with liquid. Traditional pump stage designs can handle a maximum of around 35-40% gas void fraction at the design point. However, this figure is significantly reduced at off-design conditions, severely limiting the performance envelope. Aside from the reduction in availability associated with gas locking, (requiring shut-in to re-establish prime), electric motor driven pumpsets are also at risk of motor burn-out, when the associated loss of forward flow results in a loss of motor cooling.

Extensive research into the behaviour of gas / liquid mixtures in centrifugal and axial flow impellers and diffuser resulted in the development of an axial flow stage design which avoids centrifuging of the gas / liquid fractions and maintains forward pumping. This new design, coupled with the speed response of the HSP, is effective in preventing gas locking under continuous pumping to levels in excess of 75% free gas void fraction at pump suction. Indeed the multiphase design of HSP can safely handle 100% gas slugs for short periods without losing forward flow and without suffering undue stress.

The HSP concept has design features that make it uniquely suited to handling high gas fraction fluids. Firstly, the HSP has the unique capability of varying speed automatically and instantaneously to match the pump end load. The hydraulic turbine is nominally a constant power machine, which is capable of using all of the hydraulic power supplied to it. Consequently, for any given speed, if the pump end load changes due to the mixture density changing with varying gas / liquid fractions, the HSP will respond by speeding up or slowing down to take account of the change in load. The advantage of this is that when gas slugs enter the pump suction, the HSP speeds up to compress the gas and maintain forward flow. This is a characteristic that is not available with electric motor drives, which generally operate at a nominal fixed speed set by electrical supply frequency supplied from an inverter. Although the inverter can change the speed of the ESP it cannot respond fast enough to track the load changes in slugging flow.

The second major advantage is that the HSP utilises a unique system of hydrostatic bearings to support the shaft and rotor assembly when in operation. This obviates the need for contacting bearings and rubbing liquid seals, which can be life limiting particularly when significant levels of solids are entrained within the pumped fluid. A portion of the relatively clean turbine exhaust fluid is fed to the pump end bearings and thrust balance system to provide the fluid films and keep the bearing surfaces in good condition. A further key advantage of this arrangement is that the hydrostatic bearing feed is maintained even when the pump end is handling 100% gas slugs, thus providing bearing lubrication, cooling and rotordynamic stability under all operating conditions.

The HSP has several key features that are of benefit in arduous downhole pumping environments:

• Inherent variable speed capability	Slug flow tolerant; wide, flexible operating range
• No electric motors or cabling	Eliminates potential for electrical drive failure modes
• No mechanical seals or oil barrier protectors	No dependence on components with limited life; extends MTBF
• Advanced materials used throughout	Extensive use of hard metallic alloys, ceramics and cermets mitigates wear
• Utilises clean flush to hydrostatic bearings	Non-contacting bearings free from solids abrasion and erosion
• Multiphase pumping capability	Handles high gas fractions without gas locking
• Compact, rugged assembly fully shop tested as an assembly	Typically 3-4 m in length; simple "plug-in" for deployment

All of the above factors have played a part in the successful development of the Captain Area B extension using the new variant HSP technology.

4. OPERATING EXPERIENCE

It is planned to deploy up to 15-off 20,000 bpd / 360 KW HSPs in oil producing wells accessed through the Area B subsea template. At the time of writing, a total of 9 producing wells have been drilled, equipped with HSPs and brought on stream using a semi-submersible drilling rig. The first HSP was installed and started during December 2000. On Captain an initial HSP performance test is conducted from the semi-submersible rig before hook-in to the main subsea production systems and commissioning from the surface control facilities. This test is useful as it is conducted with the well full of kill fluid of known composition and so a reasonably precise cross check can be made against the works water performance test to establish that the HSP has been installed correctly and is performing to specification.

At the time of writing reliability on all deployed HSPs has been 100%, with no instances of infant mortality, nor other premature failure events. Each production well is equipped with downhole instrumentation, which is used to monitor both well and HSP performance with time. There have been no indications of performance degradation over time for any of the HSPs deployed. Each HSP is fitted with vibration monitors in radial and axial planes to check for upwards trends indicating loss of rotor stability due to wear of critical surfaces. Measured vibration levels during steady state running have typically remained around 1g since start-up.

 C603/014/2003 © Weir Pumps Limited 2003

All of the HSPs deployed to-date have demonstrated 100% availability, with production being constrained by topsides process limitations only. There have been no instances where it has been necessary to shut-down the HSPs due to gas locking.

The HSPs have been operated over a wide flow range from zero flow, (dead headed), to 17,000 bfpd, depending on individual well productivity and operational demands.

5.0   HSP PERFORMANCE

The downhole instrumentation, together with other instrumentation systems on the subsea power water and production piping systems, has been fundamental in proving the performance of the HSPs in service. With multiphase production, (oil, gas and water), it can be very difficult to establish the true performance of pumps in the field as it is not always possible to make accurate assessments of the fluids being pumped at any particular moment in time, especially if no multiphase metering is used. In the case of the Captain HSPs, estimates of gas void fraction at the downhole venturi flowmeter have to be made, based upon production tubing losses, to enable approximate pump performance to be established and checked against the known works test performance results.

Real time relay and logging of measured data enables the offshore and onshore operating staff to monitor the performance of the HSPs and the wells and hence optimise production within the constraints of process capacity. Ready access to the data also ensures that the wells are not degraded in any way, for example by excessive drawdown through over pumping. The measured data is trend plotted for each day and can be back accessed from the database server to investigate past events and historical performance.

5.1   Multiphase performance

The ability of the HSPs to handle gassy and slug type flows has been well proven by field experience. Some of the wells are in the reservoir gas cap region and significant amounts of gas are co-produced with the oil. Gas content typically increases with time as a well is pumped, **Figure 3**. The response of the HSP is in two ways. Firstly the speed of the HSP steadily increases as the pumped fluid mixture density falls. The turbine is supplying a constant power to the pump end and the effect is to cause the HSP to speed up in order for it to use all of the power supplied. This can be seen in the speed trace in **Figure 3**, which drifts steadily upwards following the last step change in well flow with step increase in power water supply. The speed response of the HSP to fluid composition is repeatable if all other operational parameters are maintained. It can therefore be used as a means of estimating gas content.

The second effect with increasing gas content, is in response to gas slugs forming ahead of the pump in the horizontal well bore. As they arrive at the pump suction they cause the HSP to cycle up and down in speed as they pass through, **Figure 4**. With gas slugs the short term gas content at pump suction can often rise to 100%. The pump generated head falls with increasing gas content as more of the work is done in compressing the gas fraction. However, at no time does the HSP gas lock and lose the ability to generate pressure and keep the fluids moving forward. Steady state gas void fractions in excess of 65% have been pumped.

With increasing gas content the pump inlet flow increases and the pump speed increases. As the gas – liquid mixture passes through the pump the gas phase volume is reduced with some gas going back into solution and free gas being compressed. The discharge volume will depend on the compression ratio of the pump. The important feature is that the pump differential pressure, and hence forward flow, are maintained. On occasions when very high gas levels are experienced during slug flow, temporary dead-heading of the pump can occur due to insufficient pressure generation. This however is not a problem as the pump end re-establishes forward flow, without any damage being sustained. If sufficient power water system capacity margin exists then this dead-heading can be eliminated by slightly increasing the power water supply and hence HSP operating speed.

5.2   Speed and vibration response

The presence of gas within the liquid flowing from the well has a marked effect on the speed of the HSP. In slug flows the HSP shaft speed can increase by over 2000 rpm in less than 0.5 seconds. The nominal operating speed for the HSP is around 7500 rpm. However the hydrostatic bearing system maintains rotor stability between 3000 rpm and runaway speeds of around 15000 rpm, (highest recorded speed in the field so far being 15,350 rpm during start-up before the trips are armed). A high speed trip is normally set at 11000 rpm to prevent sustained operation at high speeds, which would significantly increase the erosive effects of entrained solids on the pump end components.

The shaft vibration increases markedly as gas content increases. This is due to the disturbances caused by sudden variations in fluid composition and the fact that as the liquid content is generally reduced there is less damping of the transient loads. Vibration spikes typically increase to 2-3g, although levels as high as 6g have been recorded without incident. In all cases the vibration signatures of the rotors have returned to normal levels when the HSP returns to low gas pumping indicating no adverse long-term effects.

5.3   Dead heading of the pump end

The pump discharge and the turbine exhaust commingle and flow to the surface together. However it is possible to operate the HSP with the power water circulating and with the pump end operating under closed valve conditions, i.e. dead headed. In this case the only flow leaving the pump end is the bearing feed fluid supplied from the turbine. Although not recommended for extended periods, the HSP is capable of operating under these conditions as the thrust balance and bearing systems are designed to cope. This allows operators to soft start the production wells if required by ramping them up slowly from zero flow.

5.4   Suction starvation

Another proven feature of the HSP is its ability to withstand suction starvation. On several occasions the sub-surface safety valve upstream of pump suction has been inadvertently closed whilst attempting to run the HSP, **Figure 4**. This can happen during start-ups before the low suction pressure trip is armed. In this situation the pressure within the early pump stages falls to a low level and the last few stages are effectively supporting the turbine exhaust pressure. As with dead heading, there is still a small through flow from the bearing feed system that can pass out into the pump and maintain a cooling effect.

5.5   Fluid effects

It is known that centrifugal pumping of oil / water mixtures can cause tight emulsions to form which significantly increase the friction losses and hence pumping pressure required in production systems. The multiphase stage design of the new variant HSP is predominantly axial flow in nature and so the fluid shearing and mixing action is significantly less than that of a mixed or radial type impeller system traditionally used for well pumping. The result is a reduction in the emulsion forming tendency and lower line losses downstream of the HSP. In addition, the commingled arrangement of HSP effectively doses the pumped fluid stream with power water, with the result that it is possible to put the mixture into the water continuous phase beyond the emulsion inversion point, where the mixture frictional loss behaviour is similar to that of water.

However, in determining the pump system performance we are concerned with both the downstream emulsion forming tendencies and the efficiency of the pump stages themselves in handling the produced fluids. When handling viscous heavy oils, such as those encountered in Captain, pump stage head and flow efficiency are reduced compared to that of water. The trade off required for multiphase performance is increased flow guidance through the stages, which has the effect of increasing the wetted surfaces and reducing Reynolds number within the stages. Consequently the multiphase stage efficiency is slightly less than that for a more open conventional centrifugal stage design. When the Captain HSPs are operating with dry oil the head efficiency has been slightly lower than estimated, with the result that power water used per barrel of pumped fluid has been greater. It is not yet clear why the field results have deviated from the estimated performance based upon loop tests, but the flexibility of the design and its operating envelope have ensured that there has been no production shortfall.

Further work is required to ascertain if the stage design has larger losses than estimated from the loop tests or if the Captain fluids in the field are slightly more viscous than those used in the prototype tests.

5.6   Slow speed running

The hydrostatic thrust balance and bearing system in the HSP is designed to maintain the rotor assembly in a stable condition across a wide performance range. This provides the operator with performance flexibility and removes the need for close supervision to avoid problems caused by operating beyond the bearing and shaft design limits. Stable shaft operation is only achieved above a minimum speed of around 2500 rpm where the main hydraulic thrust balancing system becomes active. At speeds below this a contacting bearing takes the hydraulic and mechanical thrust loads. This "start-up" bearing was re-engineered in ceramic following severe attrition of the previous design when subjected to frequent start-stop cycles and slow speed turning. The new bearing has performed very well and has given no indication of deterioration, despite several recorded instances of several hours running during well priming and also shut-down situations when the well has not been isolated and the rotor continues to be driven by low rate fluid flows.

5.7   HSP health checks

A series of trend plots of number groups taken from the downhole measurements are shown in **Figure 5**.

A key health check is that of turbine head efficiency, which is the ratio of turbine head drop for a given flow taken from the works tests expressed as a percentage of the same parameter on site. Any drift trend in this percentage could be indicative of an increase in the turbine swallowing capacity / bearing feed rate, which would only occur if wear was increasing the clearances in key areas. To-date there has been no adverse trends detected for the turbines indicating that good power water quality and bearing conditions are being maintained.

Another number group that is used to assess optimum HSP running conditions is that of thrust ratio. This is calculated from pump generated head divided by turbine head drop. When thrust ratio is maintained within a band from 0.55 to 0.2 the hydrostatic thrust balance system is known to be working effectively and the rotor is well supported on its main thrust bearing. If thrust ratio is too high then the pump is dead headed, whereas if thrust ratio is too low the pump end is running out and not generating sufficient head to maintain thrust balance. In general the thrust ratio has been well managed on the Captain operational HSPs.

6. CONCLUSIONS

1. The deployment of the unique multiphase HSP in the Captain field has demonstrated that high viscosity, low API crudes containing free gas in excess of 65% can be pumped successfully.

2. The HSP can maintain forward flow and significant generated head even when encountering severe gas / liquid slug flow conditions typical of horizontal extended reach wells.

3. The robust design of the HSP has shown that failure modes can be designed out of artificial lift systems to the extent that operators can become less dependant on manual control and intervention to safeguard equipment and well production.

4. Control of power water quality, produced fluid sand content and scale management are the key to extended MTBF for hydraulic drive downhole systems.

5. Downhole instrumentation is a useful tool in optimising the management of well and pump systems.

6. The multi-pump Captain development has demonstrated that HSPs are capable of being deployed 100% right first time and can provide 100% availability in difficult multiphase pumping conditions which would normally limit the up-time for conventional electric motor driven pumpsets, (ESPs).

7. HSPs have proven to be a reliable technology suitable for limited access subsea developments, where intervention to replace failed equipment and associated work over costs would have a severe impact on field OPEX economics.

REFERENCES

(1) Esson, A.L., Cohen, D.J., The development of a gas handling downhole pump, C508/026/96, IMechE Sixth European Congress on Fluid Machinery for the Oil, Petrochemical and Related Industries, Den Hague, 1996.

(2) Harden, W.G., Downie, A.A., Field Trial and Subsequent Large-Scale Deployment of a Novel Multiphase Hydraulic Submersible Pump in Texaco's Captain Field, OTC 13197, Offshore Technology Conference, Houston, 2001.

ACKNOWLEDGEMENTS

The author wishes to thank the management of Weir Pumps Ltd. and ChevronTexaco for permission to publish this paper. Special thanks go to Tom McConnell, Ronnie Graham and Angus Grant at Weir Pumps Ltd. and to Adam Downie at ChevronTexaco, for significant contribution to the preparation of this material.

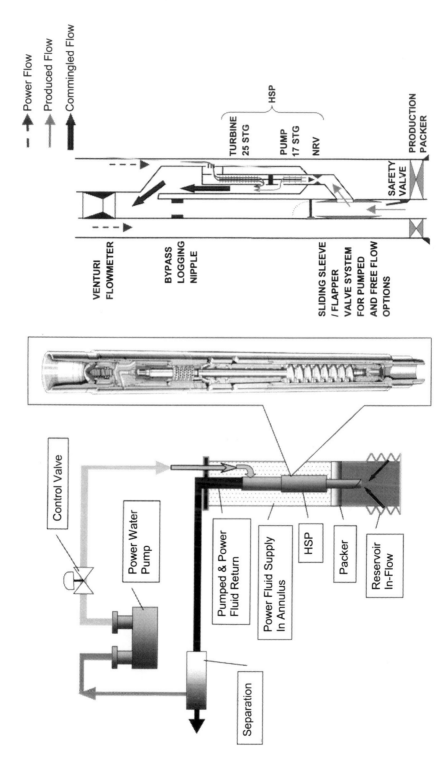

Power Flow
Produced Flow
Commingled Flow

TURBINE 25 STG
PUMP 17 STG
NRV
HSP

VENTURI FLOWMETER

BYPASS LOGGING NIPPLE

SLIDING SLEEVE / FLAPPER VALVE SYSTEM FOR PUMPED AND FREE FLOW OPTIONS

SAFETY VALVE

PRODUCTION PACKER

Figure 2. Arrangement of HSP in subsea well

Control Valve

Power Water Pump

Pumped & Power Fluid Return

Power Fluid Supply In Annulus

HSP

Packer

Reservoir In-Flow

Separation

Figure 1. HSP typical operating system schematic and sectional view

HSP In Well B2, 21 August 2001

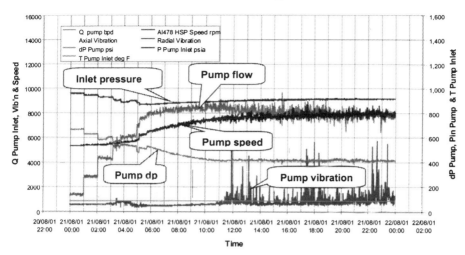

Figure 3. Response of HSP to steadily increasing gas void fraction

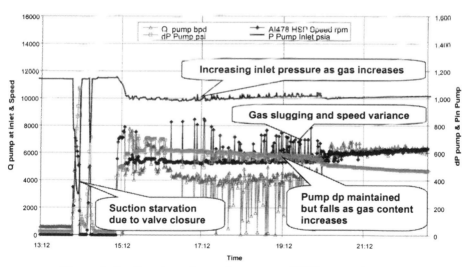

Figure 4. Suction starvation and response of HSP to gas slugs

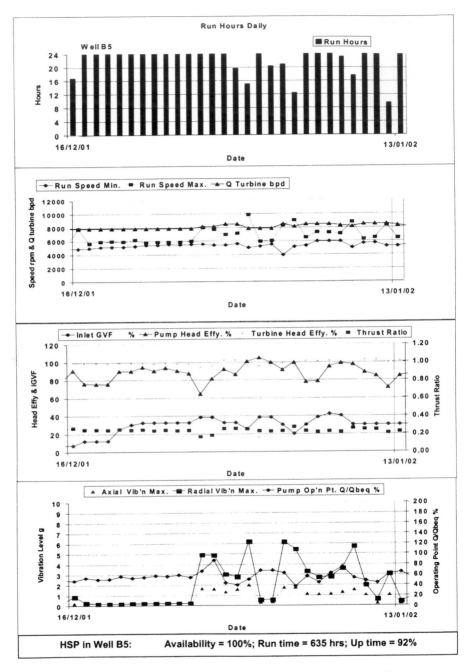

Figure 5. Typical trend plots for measured HSP performance data

C603/026/2003

Wet gas compression with twin-screw multiphase pumps

D MÜLLER-LINK, J-U BRANDT, M REICHWAGE, and G SCHRÖDER
Joh. Heinr. Bornemann GmbH, Germany

Multiphase boosting in oil and gas production may typically be characterized by:

- Slug flow occurrence possible, and
- GVF at inlet conditions higher than 50 %.

The physical challenge for the equipment in use is the strong variation of the GVF (Gas Void Fraction), leading to transient flow conditions at the pump's inlet. However, when the GVF during the entire operations constantly exceeds 98 %, control of heat generation becomes a further challenge.

Suitable answers need to be found for questions like the following:

- How can wet gas compression best be described?
- Is this an efficient way to compress wet gas?
- What about installation economics?

A technical issue for wet gas compression is the shaft sealing. Available standard seals can either cope with liquid or gas, but not with varying fractions of both. Other limiting factors are differential pressure, lubrication characteristics of pumped liquids, solids, and circumferential velocity.

1   INTRODUCTION

During the last decade multiphase boosting in oil and gas production has considerably matured. It is widely recognized as a true alternative to traditional production scenarios. The twin-screw pump is playing the major role, when the equipment has to be selected. Due to its volumetric character heavy slugging, varying water content and other typical multiphase operating challenges this pump type is well suited for this purpose. By its low speed the hydrocarbon is transported without considerable turbulence which strongly avoids emulsification whenever water is also present – a definite advantage for the later separation of the phases.

The demand for boosting an increased and more constant high Gas Volume Fraction (GVF) was only a little step ahead, then to be called "wet gas compression".

2   DEFINITIONS

Once the average gas volume fraction increases to about 98 % and only few liquid slugs, if any do occur, multiphase pumping should be called "wet gas compression". At the same time customer's interest has turned from oil to gas, and the associated liquid phase is of low viscosity, i.e. condensate and/or water. As the flow regime to the pump still is transient with changing inlet pressure and temperature, the same mechanical design guidelines must be used for the wet gas compressor as for the multiphase pump. Thermal expansion and quick temperature changes especially have to be looked after.

A typical working domain for a wet gas compressor may be described by the following set of parameters:

- Power installed: $P \leq 1.0$ MW
- Differential pressure: $\Delta p \leq 50$ bar (5 MPa)
- Total capacity: $Q_{in} \leq 1,000$ m³/h, which is equivalent to 240 MSm³/d at 10 barg inlet pressure, or 1,000 MSm³/d at 40 barg respectively,
- Inlet pressure: $p_{in} \leq 50$ barg (5 MPa).

To make the applications harder, the viscosity range is around or even below 1 mm²/s.

3   TECHNICAL ISSUES

In a twin-screw pump two pairs of intermeshing screws, forming cavities between them are rotating in the housing (see fig. 3). The fluid is thereby conveyed from inlet to discharge from both sides. Thus axial forces are fully balanced. As no metal-to-metal contact is permitted, the clearances occurring need to be sealed by liquid. The sealant characteristics increase with higher viscosities.

With little liquid occurring in the wet gas stream, it needs to be made available. This can preferably be done by storing it in the pump housing itself with a cooling circuit attached. Hence no liquid is added to the production and does not need to be separated at a later stage.

3.1   Heat generation and slug flow occurrence

Whenever liquid or gas is compressed, heat is generated. The main difference is the heat storage capability of the different fluids: good with liquid $(Water = 4.2 \, kJ/kg \cdot K)$ and very poor with gas $(\leq 1.0 \, kJ/kg \cdot K)$, the difference in mass throughput has to be observed. The discharge temperature of the fluid is governed by the pressure ratio and the fluid parameters. The latter being described with the isentropic exponent κ.

$$T_{out} = T_{in} \cdot (p_{out}/p_{in})^{(\kappa-1)/\kappa}$$

It is mainly limited by material parameters, e.g. pipeline coatings and insulation, thermal growth, etc. Fig. 1 shows the compression temperature for different pressure ratios (isentropic exponent κ = 1.2).

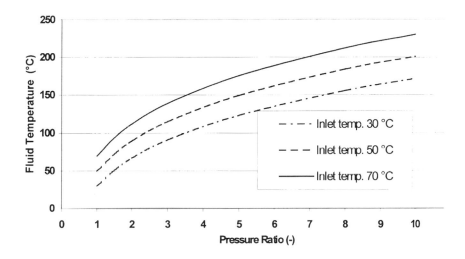

Multiphase boosting is characterized by liquid slug flow. Hence in most applications there is enough (by quantity and time) liquid available to carry heat out of the pump. In some cases of wet gas compression it may be possible, such as in gas wells filled with water. But mostly a wet gas stream does not have the necessary capability for the required heat transfer. Therefore there will be a need for cooling the liquid circulated through the pump.

3.2   Shaft sealing

It is obvious that due to differential pressures the conventional packing is not an appropriate choice for these applications. On the other hand it must be acknowledged that mechanical seals have either been developed for pure gas or pure liquid installations. But as in multiphase boosting there is also a transient flow regime in wet gas compression requiring special measures. Reference is made to a number of papers describing the challenges to be overcome (1, 2, 3). Sealing capabilities are governed by

- differential pressure across the seal,
- solids in the boosted product,
- product viscosity, and
- sealing diameter due to circumferential speed.

Table 1 shows the seal history and operating parameters for a wet gas compressor onshore Central Europe. The initial attempt saw a single acting mechanical seal failing after only 18 working hours. Besides the low viscosity of the water / condensate mixture, the seal was challenged by a high sand load from the two wells. Additionally the circumferential velocity at 1,800 rpm was 5.7 m/s. The seal had to fail.

Table 1 Wet gas compressor - seal history

Seal arrangement	Operating Conditions					Operating time
	GVF, Medium	Solids	Differential pressure	Discharge pressure	Capacity @ speed	
Single Acting Seal SiC-Si vs. Carbon g	100 %, wet natural gas	high sand content, small stones	2.0 MPa (3.8 MPa design)	6.8 Mpa g	350,000 Sm³/d @ 1,800 rpm	18 h
Smart Seal System (S3), Prototype						1 h
S3, 1st generation						97 h
S3, 2nd generation		high sand content				1,400 h
S3, 2nd generation						1,500 h
S3, 3rd generation						3,200 h, OK
S3, 4th generation						Start 02/2002

The second attempt was the installation of an unconventional new solution, the Smart Seal System (S3) invented by the company Bornemann Pumps (1). Whilst conventional mechanical seals have to provide a pressure drop and adequate sealing in one step, this is taken care of by 1. a pair of concentric throttle bushings and 2. a mechanical seal. The product leakage is routed back to the inlet of the pump, thus providing a leak-free environment. In less than one year the ongoing development matured this sealing system so as to completely fit the intended operation. More than 6,000 operating hours have been accumulated to date. Figure 2 shows a cross-sectional view of the arrangement, for more details please refer to (1).

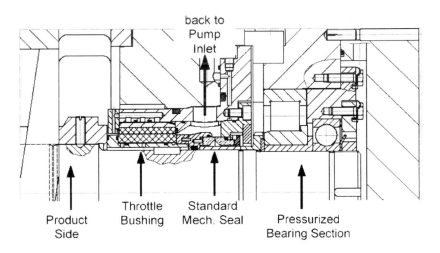

Figure 2 Smart Seal System – cross sectional view

C603/026/2003 © With Author 2003

Figure 3 shows a cross-sectional drawing of a multiphase pump in which the seal system could be implemented.

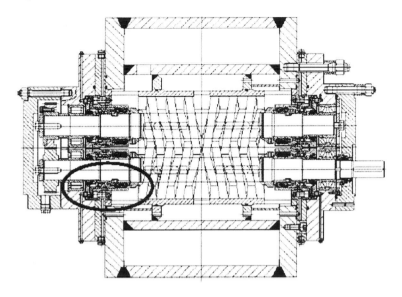

Figure 3 cross sectional drawing MPC

1   PROCESS EFFICIENCY

The rotating elements in a twin-screw pump form chambers (locks) of a constant volume. Subsequently wet gas compression has to be described as an isochoric process. The compression takes place in the locks by the slip stream due to the differential pressure, as the gas volume inside the chambers is reduced by the back flowing liquid.

This may be compared to the ideal, adiabatic process, which defines the minimum technical effort. When a twin-screw pump transfers liquid in addition to gas, the basic definition is

$$\eta_{tot} = \frac{P_{gas} + P_{liquid}}{P_{total}}.$$

Furthermore it is necessary to have a detailed look on the following losses:
- Volumetric losses, i.e. slip stream and any re-circulation of liquid, and
- Mechanical and hydraulic friction losses.

These are expressed by

$$\eta_{vol} = \frac{Q_{in}}{Q_{in} + Q_{loss}}$$

and

$$\eta_{mh} = \frac{Q_{in} \cdot (p_{out} - p_{in})}{Q_{in} \cdot (p_{out} - p_{in}) + P_{fric}}.$$

While Q names the flow and p the pressures, the indices "in" and "out" need not to be explained. Both the above are mainly dependent on the design of the machine.

Furthermore the difference between the isochoric and the adiabatic (or ideal) gas compression process is to be addressed by:

$$\eta_{ad} = \frac{P_{adiabatic}}{P_{isochoric}} = \frac{\kappa}{\kappa - 1}\left(\pi^{\kappa - 1/\kappa} - 1\right)\bigg/\left(\pi - 1\right).$$

This is only dependent on the pressure ratio $\pi = p_{out}/p_{in}$ and the isentropic exponent κ. It has to be noted that there is no influence from machine design.

With several intermediate steps the above three equations can be converted and summarized to:

$$\eta_{tot} = \frac{\alpha \cdot \eta_{ad} \cdot \eta_{vol} \cdot \eta_{mh}}{\eta_{mh} + \eta_{vol} - \eta_{mh} \cdot \eta_{vol}} + \frac{1 - \alpha}{1/\eta_{vol} + 1/\eta_{mh} - 1}$$

where α is the GVF and expressing the actual gas content at inlet conditions. The second term of the sum represents the influence of the liquid compression. It may be neglected for a high GVF ($\alpha = 1$).

As η_{mh} and η_{vol} are typically ranging from $80 - 95$ %, the total efficiency is dominated by η_{ad} (fig. 4). Wet gas compression by twin-screw multiphase pumps has its working domain in pressure ratios ranging from 2 - 5. The total efficiency then being between 35 and 60 %.

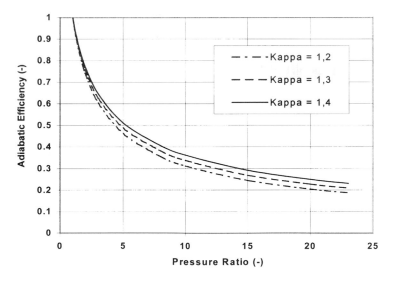

Figure 4 Adiabatic efficiency

C603/026/2003 © With Author 2003

However we are not at all looking at the efficiency of just one single machine and its efficiency, but have to consider the compression process in total, i.e. the system limits have to be set accordingly.

The end user has to value the commercial impact of a significantly simplified system lay-out. Whilst conventional compressor stations require a separator and gas drying facilities upstream and cooler downstream of the machine, there is only a single wet gas compressor installed. There is no need for a further treatment of the associated water and liquid hydrocarbons, as they are transferred to a central station. Compression heat in the wet gas compressor is transported by means of a liquid stream which is circulated through the pump and an attached much smaller cooler. Furthermore maintenance and surveillance activities are significantly reduced. This is especially favourable for remote locations.

A typical set of operating parameters for twin-screw multiphase pumps acting as wet gas compressors has been defined in chapter 2. For higher capacities a commercial evaluation of the total capital as well as operating expenditures will show, which way to go.

5   FIELD CASES

Three different field cases have been selected as examples to be presented. Table 2 shows the intended operating parameters at the start of the project. Real operating data may have been different. Whilst cases I and III are showing a relatively low inlet pressure, the pressure of case II is elevated.

Table 2 Field cases – operating parameters

Project	GVF	GLR	Liquid	Gas	Inlet pressure	Discharge pressure	Capacity	Shaft power
	%	Sm³/m³	m³/h	Sm³/h	MPa	MPa	m³/h	kW
I - onshore	97	283	30	8480	0.8	2.2	984	799
II - onshore	99.5	6895	3	20686	3.5	7.0	585	705
III - offshore	99.5	3044	6	18262	1.8	3.5	925	500

5.1   Case I

Two pumps are installed at a location in North America (4), each having a total capacity at inlet of approx. 1,000 m³/h (150 MBEPD). As required by the total flow they are operated in parallel or single respectively. The electrical drive train includes a frequency inverter. The production is delivered to a separator. From there liquid is constantly recycled to the pumps inlet in order to maintain 97 % GVF. From the operational point of view this is standard multiphase boosting with no downstream cooling required.

Both pumps are equipped with single acting mechanical seals. The installed API Plan 52 takes away the heat generated by the seal.

5.2   Case II

A gas engine drives this onshore Europe wet gas compressor. The fuel gas is directly taken from the well stream. The speed can be controlled from 1,400 to 1,800 rpm to cope with production variations. The maximum capacity of the pump is 590 m³/h or 89 MBEPD respectively.

The pump is initially filled with liquid and maintains this filling due to its patented housing design. In order to remove the compression heat, the liquid is circulated through a cooling system comprising of a liquid-to-liquid and a liquid-to-air cooler. With a discharge temperature of about 70° C, no further cooling of the delivered gas in necessary.

A high sand content during the early operation caused a big surprise. The two wells were not considered to be sand producers at all. Within one hour 10 kgs. of sand including stones of more than 5 mm in diameter were found in the strainer carrying a 500 micron filter basket.

Following the unsuccessful trial with a single acting mechanical seal, the pump was equipped with the Bornemann Smart Seal System. Necessary design improvements were accomplished within a relatively short time period (1).

5.3   Case III

This offshore installation in the North Sea of a single pump is operating downstream of a separator to boost dry gas to shore. Its total capacity is around 925 m³/h or 140 MBEPD. It is driven by a constant speed electric drive.

As in case II the liquid filling of the pump is recirculated from discharge to inlet through a cooling system reducing the temperature of the gas flow to approx. 50° C. Downstream of the pump the gas is further treated to achieve export quality.

The pump is equipped with single acting mechanical seals with an API Plan 32 flush. The clean liquid source is supplied from the produced water tank.

6   CONCLUSION

As demonstrated by three field cases, wet gas compression with twin-screw pumps is technically feasible and commercially beneficial in a reasonable operational environment. The compression heat generated is efficiently removed by liquid circulation. Shaft sealing may require special measures, but reliable systems are available. The commercial approach to wet gas compression should consider the complete system and must not overrate the sole efficiency of the wet gas compressor.

7   ACKNOWLEDGEMENTS

Substantial results of this paper, especially the ongoing development of the Smart Seal System are part of the research-project MPA (Multi-Phase Aggregate), sponsored by the German government (BMBF – Ministry of Science and Research). The authors like to express their thankfulness towards all other people involved.

8   REFERENCES

(1)   J-U Brandt, D Müller-Link, G Rohlfing (Joh. Heinr. Bornemann GmbH) "Leakage Management by a New Sealing System for Twin-Screw Multiphase Pumps" 2nd North American Conference on Multiphase Technology, Banff, Canada, 2000

(2)   N Necker, G Klier (Feodor Burgmann GmbH & Co.) D Müller-Link, J-U Brandt (Joh. Heinr. Bornemann GmbH) Rotating Sealing Technology for Twin Screw Pumps Used in Multiphase Production" 1st North American Conference on Multiphase Technology, Banff, Canada, 1998

(3)   R Maurischat (Leistritz AG) "Sealing Technology – the Key to Reliable Multiphase Pumps" 1st North American Conference on Multiphase Technology, B anff, Canada, 1998

(4)   G Wyborn (Optima Engineering) "Mobil Canada Wet Gas Multiphase Pump Project" 1st Multiphase Pump User Roundtable, Texas A&M University, Houston, TX, 1999

Safety and Environment

C603/006/2003

Development of machinery and rotating equipment integrity – safety inspection guidance notes

J J LEWIS and **B STARK**
ABB Industries, Warrington, UK

SYNOPSIS

In order to promote good practice in the design and operations of offshore installations the HSE has sponsored a suite of Safety Inspection Guidance Notes for machinery and rotating equipment. These "Notes" support non expert inspectors to understand and evaluate of the hidden risks associated with high hazard rotating equipment associated with offshore installations. They are based on a distillation of experience gained within the petrochemical industry, with the corporate knowledge of machine safety developed within ICI and ABB applied to provide this guidance. This work is reported through ABB Ltd following the transfer of the ICI Engineering Group and the associate corporate knowledge to ABB. The Project Officer for the HSE OSD Mr Prem Dua has provided technical guidance for the structure and scope of the "Notes" to suit their intended use by inspectors.

1. INTRODUCTION

The objective of the Machinery and Rotating Equipment Integrity Safety and Inspection Guidance Notes is to aid understanding of the technology used and consider aspects of the equipment, which might present hidden major risks to operators of the equipment. The health of the equipment needs to be viewed both in terms of the process integration of the equipment, and the operating and maintenance management.

As a consequence of the diversity of equipment employed on offshore installations, any inspection must cover a wide range of technology and operating requirements and provide a basis for assessment of compliance with the legislation and appropriate safety practices.

This is a real challenge. The concerns noted in the enquiry into the Piper Alpha Disaster (Reference 3) concluded the inspections carried out at that time were "superficial to the point of being little use as a test for safety on the platform", noting the limitations of sampling on the basis of "what catches the eye" within a relatively short visit to an installation, and that the guidance available for such inspections was limited.

2. BACKGROUND

Safety issues introduced by equipment for chemical and petrochemical processing plant are addressed by ICI and other Plant Operators in the design stage of an installation through the HAZOP process, and supported by a large pool of experienced Engineers. The operational safety assurance of static equipment is then verified by a registration process for pressure systems, and pressure relief devices. Historically, an equivalent process was not used for machinery and rotating equipment.

The development of a structured safety assurance and auditing approach to machinery and rotating equipment was undertaken by ICI following two significant machine failures within a short period of time over 10 years ago.

The failure of a high pressure injection pump (opposite) was one of these. This incident showed that dangers from rotating equipment may not be understood by the operators. Appreciation of the inherent dangers with equipment is necessary to ensure that equipment integration, maintenance and operation addresses such dangers.

As a result of these incidents specific learning points were identified for the need for better understanding of the safety assurance required for machinery and rotating equipment, and the opportunity was seized to address machinery safety in a more structured way.

This structured approach has been previous reported to the IMechE (see reference 1 and 2). This process was recognised by the IMechE Reliability Committee (now the Safety and Reliability Group) and through the encouragement of Bill Wong a guidance book was produced "Process Machinery – Safety and Reliability" (reference 3). This has enabled the IMechE to show how industry might respond to the needs of improvement in safety standards particularly when considered against the increasing legislative requirements from the Supply of Machinery Safety Regulations, and the Provision and Use of Work Equipment Regulations.

C603/006/2003 © ABB Industries Division 2003

The work into the understanding and auditing of operating high hazard machinery and rotating equipment safety provided a basis for establishing guidance for both the understanding and evaluation of the hazards associated with such equipment.

This work was extended within the then ICI Companies to review all the high hazard machinery in operation. Benefits resulted from the learning aspects of the retrospective reviews and the reinforcement of many generic issues for machinery and rotating equipment. The rigor imposed by the registration process has introduced an improved the documentation and effectiveness of the repair techniques practised on high hazard equipment.

The above gives a background to a process already reported through the IMechE with guidance detailed for the study process. The reality is how to take advantage of all the learning established by such a process? Obviously the awareness of such hazards help improve operations and design, but as little is documented in the standards where to turn to for such guidance?

3. DEVELOPMENT OF GUIDANCE NOTES

The guidance for inspectors provides :-

- *A top level process identifying and ranking the evidence that can be gathered on a general visit to support judgements on the apparent state of the unit.*

- *A structure for assessment of operating units by observations to allow deeper understanding of the machinery, leading to judgements on the requirements for action, or further investigation.*

- *Means by which machine related observations and auditable points, supported by additional information, can be used to evaluate the state of machine systems.*

- *Structured technical data to support the observations and assessments of the Inspector.*

Visits by safety inspection engineers to offshore installation or other process plants operating major machinery historically have concentrated on dangers due to potential contact of operators or technicians with parts of the machine. These concerns are valid, and in some cases will pose a significant immediate hazard to the operators and technicians. The contact dangers, however, are in general not the worst case event. A range of other dangers will exist from the loss of process fluid containment, loss of restraint of a high energy element within a machine, or introduced from enclosures and service supplies which can result in significant damage to the machine and anything or anybody near it.

To be effective, any inspection or review needs to be aware of the potential hazards , in terms of process and process consequential hazards, as well as mechanical hazards both visible and inherent with the equipment design.

The development of the guidance notes for the HSE OSD was structured to provide a broad appreciation of equipment, their hazards, and provide a framework for assessment.

4. APPLICATION OF GUIDANCE NOTES

> - *The Inspection Guidance Notes are split into two major parts, the Evaluation Process providing a structure for gathering and evaluating relevant information, and the Technical Guidance Notes on machinery and rotating equipment provided as a series of notes on specific systems.*

Evaluation Process

The information gained from a general visit to an installation can provide important clues to the state of the care of the rotating equipment. Inferences can be drawn from this, in some cases the need to follow up concerns is clear and can be well focused. In such cases detailed information evaluation may not be necessary.

The cultural and organisational approach to running equipment does not directly affect machine and rotating equipment safety, but taken as a whole gives the background against which a machine incident may occur. In a good operating regime the potential incident will be recognised and controlled with n o s ignificant e ffect o n t he safety o r o peration o f the p latform. In a p oor operating regime an incident may reach a dangerous state before its effect is recognised.

The challenge for any inspection or audit of an installation is to identify the operations that are being well done, those that will need development, and a ny practices that may l ead t o serious incidents. The gathering of the right information and the evaluation of it, supported by appropriate guidance needs to be structured.

The map shows the structure for a process which an Inspector goes through during a general visit. The guidance documents have been prepared for both training and reference purposes.. These identify topics for review and associated evidence of a satisfactory system. They have been split into 3 streams: Induction / Meetings ; Control Room ; Plant Tour.

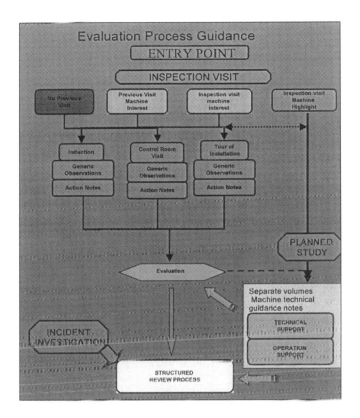

The observations made of the general state of the installation and manning is "filtered" through an evaluation matrix, with a view to identifying those issues / practices which raise sufficient concern to justify deeper investigation. Any particularly serious concerns may require immediate discussion with the OIM.

The "filtered" observations can then be used to aid in the planning of a structured review of those parts of the installation which give rise to concerns. The approach may equally be used to deal with non-machines issues.

The Guide provides two fundamental levels of evaluation for the assessment of information gained during initial or general contact with a facility.

Use of Word Models
Firstly at a summary level, where general impressions have been reached from experience or by the limited nature of the information available

For simplicity, a "3 row" word model is used, based on the activity to be performed providing examples and concerns against key words.

The evaluation guidance is structured allowing a summary level approach, a detailed approach, or a combined approach, with consideration of the detail where required.

The results from the evaluation of elements is then focused on follow up activities by considering the synergies between areas of observation. This allows the observations made during initial contacts to be melded with technical background on the equipment under consideration to aid the development of a strategy for further investigation.

Evaluation Summary

The process is summarised by relating the points of concern to the structured review process. In practice the relationships between observations and evaluations are handled on a database, with the chart indicating the underlying structure for the process.

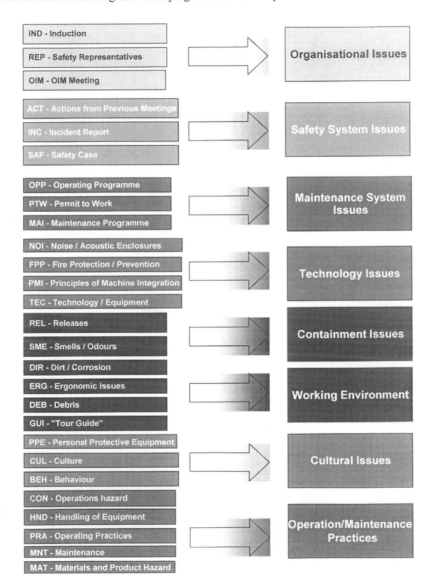

C603/006/2003 © ABB Industries Division 2003

5. STRUCTURED REVIEWS

- *The objectives of the structured review are : -*

- *To confirm that engineering equipment and associated systems are designed, installed and maintained in accordance with legislative requirements, good engineering practice and in a manner that ensures that they are fit for purpose for safe operation for a defined period.*

- *To ascertain compliance with statutory requirements.*

- *To ascertain that engineering teams are working to defined procedures, best practice, or appropriate relevant Site instructions.*

- *To raise awareness of the necessary engineering knowledge, standards, best practice and to encourage staff to seek appropriate technical help, advice and training.*

The evaluation process described above allows the data gathered during any inspection or audit visit to be considered against the needs for of the different equipment using the information in the Technical Guidance Sections to support judgements made.

The quality of sample auditing was questioned in the report on the Piper Alpha Disaster (Ref 3), to avoid such auditing being superficial it is important to have structured processes to ensure the quality of the review and knowledgeable people available to carry out such audits.
The guidance provides a framework for reviewing the procedures, processes and practices on a facility. The approach gives an overall indication of the context, desired standards, and effectiveness of activities. Evidence gathered during a structured review provides the basis for technical consideration of the facility and support for the conclusions reached by such analysis. It also gives the opportunity to probe sensitive areas to show the adequacy or otherwise of the systems, practices, and equipment.

The judgement on the systems relating to machinery and rotating equipment is supported by the technical reviews (described later) within the Guidance Notes, which can be used both as preparatory support and evaluation of the information found.

During the course of such studies the effectiveness of organisation practice can be reflected back to the Operating Management. Even straight forward observations on the availability of information can be significant. In one case it was found that the quality procedures ensued that the operating team had access to all current procedures , but only in the control room. Maintenance was carried out well away from the control room and consequentially suffered from the lack access to the maintenance procedures and information necessary to carry out the repairs.

6. TECHNICAL SUPPORT GUIDANCE

Technical Guidance Notes
The technical guidance is written in a series of sections covering packaged machine systems. There are separate notes giving more detailed information on the specific machine and rotating equipment included in the package and the ancillary equipment installed to support the operation of the packaged equipment.

The technical guidance covers :
- General Description
- Main components
- Main sub systems (seal supply, lubrication)
- Safety systems
- Main services

with identification of :-
- Hazards
- Operation
- Maintenance
- Control
- Key technical areas

The technical guidance gives both an outline of how the equipment should work and in contrast how it responds when things go wrong, and how they might fail in ways which would give concern for safety. These comments have been drawn largely from the experience within ICI where a systematic process had been established to identify and share the learning from failures. The series of guidance notes covers common packages of equipment, which are supported by more detailed notes on the major equipment items.

Treatment of Equipment within Inspection Guidance Notes
The guidance notes provide information on the machines in terms of background and history, hazards, operational needs, maintenance, and identification of the main components.

In this way the guidance helps the understanding of the task of the machine and provides an overview of the issues that can be experience from operation and maintenance of such equipment.

The technical guidance notes have been produced to cover 15 typical machine package with detail on the machines and rotating equipment within each package as well as the ancillary systems and equipment.

The notes show where learning from failures of similar equipment can be used to anticipate problems with currently operating equipment. These notes prompt questions about the current operating or maintenance practice to establish how these potential pitfalls can be avoided. The guidance has been developed to cover most of the packaged machinery equipment present on offshore installation and brings together information on how the equipment should operate as well as what happens when things go wrong.

For example, in the case of the reciprocating compressor on a flammable gas duty the range of hazards are considered. The assessment needs to address the issues from the operational access to shut down the equipment through to the consequences of liquid ingestion. The latter may lead to the machine being wrecked. The guidance notes allows users and reviewers to appreciate the effect of machine deterioration.

For a reciprocating compressor this could be from the machine valves passing causing high temperatures, interstage pressure, loss of forward flow, and allows comparison with other issues such as liquid ingestion which can cause devastating effects.

Learning points can be drawn from examples such as the consequence of failing to fully secure and check distance piece bolts to the cylinder head. In this case these have progressively loosened and lead to the cylinder head displacement, with the consequential miss alignment causing the cross head to fracture. The machine has then run on with extreme vibration levels until security of all the bolted connections is compromised. Apart from the immediate root cause to ensure that the maintenance is done correctly, further learning can be deduced about the value of high vibration trip systems, and even the position of the emergency stop control.

A range of pumps is also covered within the guidance, again the hazards associated with standard pumps are recognised and then for particular types these are developed. For multi-stage pumps operational hazards are extended to cover:

Multi-stage pumps have a low tolerance to a dead head pumping situation with a consequent rapid rise of internal temperatures and pressures.
Discharge pressures can easily exceed 100 barg, even a pin-hole leak can give a dangerous fluid jet or atomised spray of considerable length.
Higher head rise increases the sensitivity of the design to changes in fluid density, the design should accommodate all operating conditions – start up, normal operations, operating excursions, washing / purging operations, and shut down.

These learning points can be presented in such guidance to show the importance of the attention to detail necessary for all machine maintenance. As failure of machines can be identified by monitoring, and avoided by use of appropriate protective devices, the amount and effect of monitoring, and the protection requirements need to be identified. For example operational monitoring:

Operational Monitoring	Gas Turbine	Centrifugal Compressor	Recip Compressor	Pumps	Motors	Diesel Engines
Inlet / outlet temps	Y	Y	Y			
Inlet / outlet pressure		Y	Y	Y		
Power draw / fuel flow	Y	Y	Y	Y	Y	Y
Flow forward / recycle		Y	Y	Y		
Minimum flow	Y	Y	Y	Y		
Process density / MW		Y	Y	Y		
Speed	Y	Y		Y		Y
Process feed density / Mole Weight / contaminants		Y	Y	Y		
Process feed availability		Y	Y	Y		

The observation that is relevant to inspection visit are described in word models where impact , inference and possible actions are shown an extract of this for a reciprocating compressor:-

	TOPIC	KEY OBSERVATION	IMPACT	INFERENCE	ACTION
Equipment	Vibration	High vibration levels, particularly if unsteady	Machine damage. Compressor trip. High vibration levels can cause loosening of machine and connection bolting leading to accelerating collateral damage. Vibration can result in fatigue damage to connections.	There is damage to the motion works. Loosening or fatigue of process connection may result in a significant emission.	Carry out frequency analysis of vibration signals. Identify cause and plan remedial action.
	Flexibles	Flexible connections	Potential failure	Flexible connections are easily damaged	If a flexible connection is necessary, it should be of an appropriate grade and on an inspection register.
	Bearing temperatures	Bearing temperatures high (alarms in, check temperature instruments)	Bearing failure. Machine trip Mechanical damage (as consequence). Process gas release.	Bearing or lubrication fault has not been identified by operators	Root cause analysis, appropriate remedial action
	Noisy compressor valves	Clattering can be heard and felt local to valve(s)	Loss in performance. Valve could fail or damage valve cover.	Valves are damaged or loose.	Monitor vibration at each valve, identify severity, trend to check for deterioration.
	Baseframe	Machine feet moving on bed (Cracked paint or "panting" - movement)	Misalignment. Possible major failure.	Holding down bolts loose or broken	Tighten bolts. Check freedom of sliding connections. Check for added constraints which should not be there
	Oil leakage	Oil leakage or visible plume from oil system vent.	Loss of oil. Process gas in crankcase.	Check for damaged oil pipe. Oil fume could indicate a failed rod packing.	Monitor closely. Identify fluid and source. Remove source of pressure. Repair.

The process for evaluation outlined above allows the methodology to be written into an expert system, where the evaluation can be established and supported by a series of structured questions. Currently this has been done for a limited amount of the process and further development will depend on the final requirements to provide an interactive system for the users.

ACKNOWLEDGEMENTS

This work has been sponsored by HSE OSD with Mr Prem Dua as the Project Officer. The writing of the document by the authors has been supported and enhanced by their colleagues in ABB Ltd.

7. REFERENCES

1   C449/013 IMechE 1993 by Dr Harald Carrick and Mr Keith Rayner
2   Registration and Verification of Critical Machines Mr J J Lewis – IMechE Seminar Maximising Rotating Reliability Dec 94
3   The Public Enquiry into the Piper Alpha Disaster (Cm 1310) – The Hon Lord Cullen
4   "Process Machinery – Safety and Reliability" Editor Bill Wong (ISBN 1_86058_046_7)

C603/021/2003

Improving safety at gas turbine plant – preparation for ATEX

I R COWAN and S GILHAM
WS Atkins Consultants, Epsom, UK
E S KAUFMAN, R L BROOKS, and L M DANNER
GE Power Systems, Greenville, South Carolina, USA

ABSTRACT

WS Atkins (WSA) has been working with General Electric Power Systems (GEPS) for over five years to advance the safety and performance of their gas turbine enclosure ventilation and hazardous gas leak detection systems. Initially, this work was in response to regulatory changes in the UK power industry, prompted by the UK Health & Safety Executive (HSE). However, GEPS have pursued the incorporation of the HSE philosophy in the development of hazardous gas protection across the GE product line. By using the UK HSE dilution ventilation criterion to generate a basis for safety, GEPS are ensuring that they are ready to meet the demands that the forthcoming ATEX Directives will place on the manufacturers of gas turbine equipment. This paper reviews the safety regulations with regards to gas turbine enclosures, and illustrates how we have devised an integrated approach to safety and performance, based on effective ventilation and gas detection.

1   INTRODUCTION

The last decade have seen a considerable worldwide increase in the number of gas turbine based power plants. Approximately 31% of the UK's power is generated from gas, compared to 1% a decade ago. Gas turbine power plants can offer a combination of high efficiency (up to 60%, when operated in combined cycle) and low emissions, are relatively quick to construct, require less capital cost, and have in the past benefited from competitive fuel prices.

Gas turbines are typically housed in an acoustic enclosure in order to protect the turbine and to eliminate environmental noise (see Figure 1 for an example installation). However, in the event of a gas leak, such enclosures can lead to a build up of gas and thus give rise to an explosion risk. Most gas turbine enclosures are fitted with ventilation and gas detection systems, and rely on these systems to provide a basis for safety against the explosion hazard. Alternative approaches, such as the inclusion of explosion relief panels or explosion suppression systems, are also used for certain installations but are typically not appropriate for onshore power plant (Ref. 1).

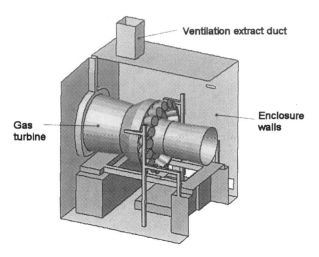

Figure 1. Schematic of a GEPS 9FA gas turbine enclosure (two of the walls and half of the roof have been cut away to show the contents of the enclosure).

The purpose of the ventilation and gas detection systems is to prevent the undetected build up of a hazardous gas cloud, as follows. The detection system is designed to reliably sense the presence of large leaks, which could cause a significant explosion hazard. The ventilation system is designed to ensure that leaks that are too small to be detected are rapidly diluted, thereby preventing hazardous gas build up. It is also designed to ensure that large (detectable) leaks are purged rapidly from the compartment, once the leak ceases. 100% redundancy is employed for both the gas detector sensors and the fans, in order that failure of one of the components does not compromise the safety of the system.

This approach recognises that it is not practicable to design the ventilation system to fully dilute the large leak rate events, such as a full bore rupture of a fuel line, which have an extremely low probability of occurrence. Instead the gas detection system is designed so that an alarm provides sufficient warning of hazardous gas leaks, prompting removal of the leak source. The introduction of this "dilution ventilation" basis for safety for gas turbine enclosures has been relatively recent (i.e. in the past five years), driven partly by regulatory pressures in the UK. However, the forthcoming implementation of the ATEX directive will require this or an alternative basis for safety to be present for each plant installed in Europe.

This paper reviews how we have enhanced product safety of our plant by developing a dilution ventilation system for protection against hazardous gas leaks in gas turbine power plants. Much of the discussion is directed towards gas turbine enclosures, but the same approach is also used for other enclosures containing flammable gas systems, for example gas valve modules and hydrogen-cooled generator collector compartments. Similar analysis techniques have also been applied to study fire detection and fire protection systems. The above advances have been accompanied by improvements to the engine performance, aided by the same analysis techniques that were employed in studying the safety issues. An example is the introduction of a clearance control system on GEPS's H-series turbines, for which the ventilation design has a significant impact.

2   EVOLUTION OF SAFETY STANDARDS FOR GAS TURBINE PLANT

Prior to 1996, there was no consensus on the definition of the ventilation rate that was required to ensure sufficient dilution of accidental gas leaks. There are a number of area classification, fire protection, and purchasing standards and codes that are relevant to gas turbines (e.g. IEC79-10, the Institute of Petroleum Code IP15, the NFPA Code 37), some of which suggest the use of ventilation as a basis for safety. Reference 1 gives a detailed summary of all of the relevant standards.

One such code is IEC79-10 (Ref. 2), a European hazardous area classification standard for the design of electrical apparatus in explosive atmospheres. The standard provides an expression for calculating the hazardous cloud size, based on the leak rate, ventilation flow rate and flow "quality". This can be written as:

$$V_{pock} = f Q_{min} / q$$

where V_{pock} is the gas pocket size (in m^3), f is the ventilation quality factor (ranging from 1 for perfect unobstructed flow, to 5 for strongly obstructed flow), q is the ventilation flow rate (in air changes per second), $Q_{min} = Q_{leak} / (50\% C_{LEL})$ is the flow rate of air required to dilute the gas leak down to a concentration of 50% of the lower explosive limit, C_{LEL}, and Q_{leak} is the gas leak rate (in m^3/s). This empirical expression is based on a simple mixing model, and does not take account of the self-induced mixing that occurs as a result of the high speed jet release of the high pressure fuel.

The gas pocket size from the above expression can be compared with the size of the enclosure to assess whether the gas cloud is negligible ("high degree of ventilation"), or is significant but contained within the enclosure ("medium degree of ventilation"). For typical enclosures, the former would lead to a non-hazardous classification, and the latter to a Zone 2 classification.

Taking a typical GE enclosure (9FA engine) as an example, this calculation would suggest that flow rates of approximately 40 and 40,000 air changes per hour (ach) would be required to achieve medium and high degrees of ventilation respectively, based on "poor quality" flow and a typical minor flange leak, i.e. a 20g/s partial section leak from a flanged joint (see Ref. 3 for a definition of this leak event, and for data on leak frequency). The latter figure is two orders of magnitude higher than is feasible, and demonstrates that only a "medium degree of ventilation" can be achieved in practice, according to the above expression for gas pocket size. This is consistent with the Zone 2 classification rating that is typically employed for turbine enclosures.

In practice, a medium degree of ventilation is easily achieved. Indeed, ventilation rates of 80ach or higher are usually employed, in order to maintain an acceptable thermal environment inside the turbine enclosure, and in order to ensure that the compartment is purged rapidly of gas following large scale leaks or the activation of the carbon dioxide fire suppression system. Hence, with this approach, the main driver for the ventilation design is not necessarily the adequate dilution of gas leaks, since this requirement is apparently easily met. This may be the reason why, historically, manufacturers have not focused detailed attention on the design of the ventilation and gas detection systems.

In 1997, the UK HSE issued an interim advice note (Ref. 4) to the UK power industry, in response to a number of explosion incidents and to investigations by WSA and the HSE into gas build up in typical turbine enclosures. This note covered all aspects of health and safety at Combined Heat and Power (CHP) and Combined Cycle Gas Turbine (CCGT) power plant, and focused in particular on the potential for gas explosions in the gas turbine enclosure.

A major development in the advice note was the drawing of a distinction between "adequate dilution" and "dilution ventilation" for undetectable leaks, and the introduction of a quantitative criterion for the latter. In IEC79-10 terminology, these two regimes were made equivalent to medium and high degrees of ventilation respectively. The ventilation rate is "adequate" if it controls the gas cloud size to a volume less than the enclosure bounds. Under these conditions a sizeable gas cloud can be present, and so since the ignition probability is understood to be high (another key development) and the leak duration is likely to be prolonged, this can lead to an unacceptable explosion risk.

Dilution ventilation, on the other hand, requires that the gas cloud be diluted rapidly, leading to a small gas pocket size, so that ignition of the gas does not generate a hazardous overpressure. A criterion was presented in the advice note stating that dilution from pure gas to 50%LEL must occur within a volume of 0.1% of the net volume of the enclosure, for a leak which would just trigger gas detection. (LEL denotes the lower explosive limit, as a volume fraction of gas, i.e. approximately 4.4% for natural gas in air). This criterion is based on the assumption that ignition of such a cloud would result in an over-pressure inside the enclosure of approximately 10mbar, which is a typical enclosure strength. Gas clouds larger than this are expected to cause a failure in the enclosure, with consequent hazard for nearby personnel and equipment.

The dilution ventilation criterion is based on a simple empirical estimate of the critical pocket size, and contains several levels of conservatism. For example, the gas pocket size is based on the volume of gas at a concentration of 50% of the lower flammability level and higher. Much of this gas may thus be non-flammable. However, this currently remains the best criterion for the critical gas pocket size.

Earlier in the Section, it was demonstrated that a rapid dilution is not feasible for typical leak sizes, according to the usual empirical expressions for gas pocket size. However, computer simulations by WSA have found that these expressions are not appropriate for most gas leaks, since they ignore the effect of the momentum in the leaked gas, and the mixing that this induces. The fuel gas systems operate at pressures of the order of 15bar or higher, so that any leaked gas will be realised as a high-speed, under-expanded jet. This jet induces significant local air movement and mixing, and will aid dilution of the leaked gas provided that the air entrained into the jet is free from gas (the presence of stagnant or recirculating flow regions can lead to a high background level of gas, which will inhibit this self-dilution effect). Hence, if the background ventilation flow does not contain areas of stagnant or recirculating flow, and if the leak jet does not impinge on a surface close to the leak source, then the leaked gas jet can effectively self-dilute. This means that the resulting gas pocket size may be considerably smaller than that suggested by the earlier simple expression, and so dilution ventilation can be a practicable design goal.

The HSE found that many of the existing UK installations failed to meet their criterion for dilution ventilation of undetectable leaks, either because the alarm settings for the gas detectors were set to high values to prevent spurious alarms, or because the flow patterns inside the enclosure included stagnant or recirculating flow regions which promoted gas build up. In particular, there were a range of gas leak sizes that would remain undetected but,

C603/021/2003 © IMechE 2003

because of poor mixing within the compartment, could form a substantial gas cloud within the compartment. For these installations, the basis for safety was impaired.

The HSE advice note generated an industry-wide review in the UK of the safety systems at gas turbine plants, and prompted gas turbine manufacturers to review their own procedures and designs. In July 2000, the HSE followed up with a guidance note, PM84 (Ref. 5), which provided additional advice on identifying hazards at gas turbine plants and reducing fire and explosion risks. In particular, the guidance note recognised that a combination of computer modelling and field testing was required to analyse the performance of the ventilation and gas detection systems, and to ensure that dilution ventilation was achieved for undetectable leaks.

Finally, from July 2003 gas turbine plants in Europe will be required to comply with the new ATEX Directives (94/9/EC and 1999/92/EC), which are concerned with the use of equipment in potentially explosive atmospheres. These directives are aimed at the free movement of goods across the common European market, and as such aim to harmonise safety standards. Directive 94/9/EC states in Annex II that "potential ignition sources such as ..., high surface temperatures, ... must not occur", which appears to cause a problem for some gas turbine hot surfaces. However, it is recognised that the gas turbine forms part of an integrated package which can be shown to comply with the directive, even if the individual components (i.e. the turbine) are not compliant. In effect, this enables the installation of a potential ignition source inside a Zone 2 hazardous area by considering the turbine, its enclosure and its associated safety systems as an integrated package. Adoption of the above dilution ventilation basis for safety, in association with appropriate operational procedures, has been accepted by the European Commission as one approach to demonstrating compliance with the ATEX Directives (Ref. 6).

3   DILUTION VENTILATION AS A BASIS FOR SAFETY

GEPS have adopted dilution ventilation as the basis for safety for their gas turbine enclosures. This relies upon two factors: good quality air flow within the enclosure; and reliable gas detection. The former ensures that leaks that are too small to be detected are sufficiently diluted, whilst that latter ensures that detection will occur for those leaks that do lead to an explosion hazard. These issues are now explored in more detail. A discussion is also presented on the effects of fuel ingestion into the enclosure inlets and of liquid fuel leaks.

3.1   Flow quality

A major aim of the ventilation system design is to achieve good flow quality within the gas turbine enclosure. This means that regions of flow stagnation and recirculation should not be present in the vicinity of any likely leak sources. Note that leaks from high pressure fuel lines have the momentum to travel several meters, as demonstrated by the data in Table 1, so that this vicinity can include a substantial proportion of the enclosure.

Computational Fluid Dynamics (CFD) is an effective tool that allows simulation of the flow and heat transfer within the whole enclosure, thereby providing a three-dimensional picture of the conditions within the compartment. This can be used to identify regions of poor flow, and to test different design options for the ventilation system. Design changes might involve minor modifications to the location or size of the ventilation inlets, or a complete redesign of the entire system. For example, we have used the latter approach to design a clearance control ventilation system that is aimed at significantly improving the gas turbine operational efficiency, as well as providing dilution ventilation for gas leaks.

Leak source size (mm^2)	Leak rate (g/s)	Distance from source at which the jet velocity drops to …	
		5m/s	1m/s
0.44	1	0.8 m	3.7 m
2.3	5	1.8 m	8.5 m
4.6	10	2.5 m	12.0 m
9.2	20	3.5 m	17.0 m

Table 1. Details of the evolution of a round jet into a free atmosphere, for four different sized leaks from a 15bar methane fuel line.

Once a design has been chosen, then gas leaks into the enclosure can be simulated in order to demonstrate the effectiveness of the final design (this is discussed in more detail in the next Section). Further redesign of the ventilation system may then need to occur if the leak simulations find that the gas dilution is not sufficient. Note that these leak simulations automatically include effects such as the self-dilution of the jet due to its own momentum, and the interaction of the jet with stagnant or recirculating regions and with solid surfaces within the enclosure.

Fuller details of WSA's analysis approach are provided in an earlier paper (Ref. 7). In particular, this paper stresses the importance of supporting such computational work with field tests, in order to ensure that accurate boundary conditions are applied in the CFD model, and to enable a level of validation to be undertaken on the simulations. It remains a considerable challenge to the engineer to generate a detailed CFD model of the enclosure, which contains complex geometry and boundary conditions. Assumptions and simplifications are inevitably made, and it is important to be able to assess the effect of these assumptions through comparison with field data.

The simple expression presented in Section 2 has an inverse relationship between the hazardous gas pocket size and the ventilation flow rate ($V_{pock} \propto 1 / q$). This relationship suggests that the optimum approach to gas dilution is to employ the highest ventilation flow rate that is practicable. There are two problems with this approach, however: firstly, gas leaks become progressively harder to detect as the flow rate increases; and secondly the increases in flow velocity within the enclosure are likely to be small, and therefore have only a minor effect on the gas dilution for a given leak size.

The first point can be demonstrated by considering the gas detector setting that would be required to ensure detection of a leak that generates a critical gas pocket size, for different flow rates. Assume first that a leak rate of 19g/s is the critical leak size, above which an excessive gas pocket is produced. Assume also that this leak rate is not significantly affected by an increase in the flow rate (this is discussed next). Then, a doubling of the ventilation flow rate will halve the gas concentration at the detector, so that a set point of half the previous value will be required in order that the system alarms for this critical condition.

C603/021/2003 © IMechE 2003

To illustrate the second point, consider the typical ventilation flow velocity inside the enclosure: with a flow rate of q = 80ach = 0.022s^{-1} supplied at the bottom of the enclosure and extracted at the top, and a typical enclosure height of H = 10m, then the average flow velocity inside the enclosure is calculated as:

$$U = Q/A = qV/A = qH = 0.022 * 10 = 0.2 \text{m/s}$$

where A is the compartment cross section and H its height. Hence, even a fivefold increase in ventilation flow rate would only produce an average flow velocity of 1m/s. This velocity is typically much lower than that in the leaked gas jet, and so does not have a significant impact on the initial evolution of the jet.

Furthermore, stagnant or recirculating flow regions present in the flow are driven by the geometry of the enclosure and the design of the ventilation supply and extraction. These regions will not be removed by increasing the ventilation flow rate. Instead, the appropriate action is to redesign the ventilation supply, in order to sweep these regions more effectively.

Hence, our design approach has been to apply the minimum flow rate that can be accommodated from a thermal design viewpoint whilst still maintaining good flow patterns within the enclosure, diluting minor gas leaks, and purging large gas leaks in an acceptable time. The latter is achieved through careful consideration of the location and design of the air supply and extraction. This approach delivers the optimum performance from the gas detection system.

3.2   Gas detection

There are two main issues regarding gas detection: choice of appropriate alarm and shutdown settings, and reliability of the detection. These two issues are discussed in turn.

3.2.1   Detection reliability

It is important to site the gas detectors appropriately, in order to ensure that the detectors are reliable in sensing the presence of gas leaks inside the enclosure. This means that gas should pass over the detector, wherever the gas leak occurs. Typically, this means siting the main detectors in the ventilation extraction duct(s). Additional detectors may be used inside the compartment to provide early warning of certain leaks into certain regions and to help diagnose leak locations, but these cannot be relied upon for all leak scenarios, since they can miss leaks from locations in other parts of the enclosure.

Our approach is to employ a redundancy-based design using a multiple sensor array. This means that one of the gas detector sensors can fail without compromising the safety of the system and thus requiring shutdown of the plant. This is discussed in more detail in Section 4.

Finally, it is important to ensure that the air flow over the detector array is thoroughly mixed, in order that the concentration in the flow passing over the sensors is representative of the flow passing out of the enclosure. CFD simulations can again be used to address this, and to design additional mixing devices should they be required.

3.2.2   Detector settings

Choice of the set points for the gas detectors is a compromise between achieving the highest accuracy without generating regular spurious alarms or trips. GEPS employ catalytic bead detectors, whose current accuracy is approximately ±3%LEL with a minimum practicable set point of approximately 5%LEL. The chosen set points must take into account these sensor accuracy bounds.

We define two set point levels for the detectors. The first level is based on a gas cloud defined by the 50%LEL gas concentration (HSE dilution ventilation criterion) and is in the range of 5% to 10% LEL. The second level, in the range of 10% to 25% LEL, will trip the gas turbine, and is based on a gas cloud defined by the 100%LEL gas concentration. More discussion is devoted to GEPS's design practice in Section 4.

The choice of the detector settings is based on results from a series of gas leak simulations with varying leak rates, all directed into the worst case location. For each simulation, a gas pocket size is calculated (according to either the 50%LEL or 100%LEL gas concentration) and plotted against the gas concentration that this would generate at the detectors in the ventilation extraction duct. An example curve for setting the first level of detection is shown in Figure 2. Due to the strong non-linearity in the curve, it is usually found that at least three gas leak simulations are required, with data points that straddle either side of the criterion value.

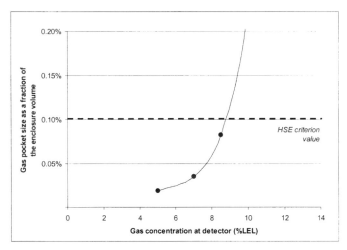

Figure 2. Example of the results from a series of gas leak simulations into the worst case location for gas build-up within a GEPS enclosure.

By noting where the curve crosses the HSE criterion value for the pocket size (0.1% of the enclosure volume), a first estimate can be made of the required alarm setting. In the example curve, a setting of 9%LEL would be estimated. Allowance is then made for the accuracy of the sensor — for example, a ±3%LEL error bound on a set point of 9%LEL would generate alarms for gas pocket sizes of 0.03% to 0.5%. The latter exceeds the gas pocket criterion by more than a factor of two, due to the steep nature of the curve at that point, and so is probably not acceptable. A set point of 6%LEL however would alarm for pockets in the range 0.02% to 0.1%, and so is a better choice.

  C603/021/2003 © IMechE 2003

3.3 Fuel ingestion through ventilation inlets

The previous discussion has dealt with fuel leaks inside an enclosure. However, the enclosure ventilation system source is air from the surrounding environment, therefore leaking fuel lines or components outside the enclosure can lead to fuel ingestion and potential gas build-up within the enclosure. Enclosures with a single ventilation inlet are particularly susceptible to this hazard, since the air inside the enclosure will be gradually replaced by fuel, at the average concentration ingested at the inlet. A simple model for this is:

$$C / C_{in} = 1 - \exp(-q \, t / f)$$

where C is the average concentration inside the enclosure, C_{in} is the concentration at the ventilation inlet, q is the number or air changes per second through the enclosure, f is the ventilation flow quality (value from 1 to 5 — see Section 2), and t is the time from the start of the ingestion. This expression can be rewritten to give the time taken for the concentration in the enclosure to rise to a given value, C_1:

$$T_1 = (-f / q) \ln\{1 - C_1/C_{in}\}$$

Using GEPS's 7EA gas module as an example, for which $q = 75$ach $- 0.0208$ changes per second, $f = 1.1$ (the enclosure is well mixed), and $C_1/C_{in} = 50\%$, then the above expression predicts that it takes only 37 seconds for the concentration in the enclosure to rise to 50% of the value at the inlet. Within three minutes, the concentration in the enclosure is within 3% of the inlet concentration. These predictions were validated by a CFD study.

The above analysis has thus shown that a potential gas build-up hazard can exist inside a single-inlet enclosure if there is a prolonged ingestion of fuel at a concentration in excess of the fuel lower explosive limit (LEL). This hazard can be detected by the gas detection system within the enclosure, if one is installed. As a result of these findings, we have developed a PC-based tool for predicting the probability of hazardous fuel ingestion into a single inlet enclosure from a network of piping. The tool is statistical in nature, and is based on empirical models that calculate the distance required for a fuel leak to dilute down to non-hazardous concentrations. The tool caters to both gaseous and liquid fuels.

Finally, for enclosures with multiple inlets, fuel ingestion would typically not occur through all the inlets simultaneously, since the inlets are typically placed on opposite sides of the enclosure. Hence, it is unlikely that hazardous gas build-up would occur throughout the enclosure. There is, however, a potential for significant gas build-up near the affected inlet(s), and the above tool can be used to evaluate these situations.

3.4 Liquid fuel leaks

Liquid fuel systems are used in many gas turbines, and operate at high pressure. A leak from a liquid fuel line will form an atomised liquid spray (or "mist") which can pose a fire hazard, even though the fuel temperature is well below its flash point. If substantial build-up of the fuel mist occurs within the enclosure, then an explosion can result with the same consequences for safety as a gas explosion (see for example Ref. 8). This has occurred in the UK for a unit operating on Naphtha.

Currently, there does not appear to be a practical fuel mist detection system that can provide the equivalent protection of gas detector systems. Systems do exist that will detect the presence of a mist, but these can also be triggered by water mist and dust. Hence, the ventilation dilution approach for protection against gas leaks cannot be applied to liquid fuel leaks with a high degree of confidence due to the lack of reliable detection.

Instead, we identify the potential hazardous range of liquid fuel leaks and ensure first that potential ignition sources are removed form the hazardous area and second that ventilation inlets are protected from contamination by leaked fuels. This approach minimizes the potential for explosions from this source until effective detection technologies can be developed to enable an effective fuel mist dilution ventilation approach.

4   GEPS DESIGN PRACTICES

The following discussion illustrates how the dilution ventilation basis for safety has been incorporated into the GEPS design practices. To develop an appropriate strategy for protection against hazardous gas, we employ an integrated hazardous gas protection methodology to fully optimise the ventilation system in conjunction with hazardous gas detectors and control logic. The system strategy warns operators when gas leaks are not sufficiently diluted and further provides a tripping feature, which will automatically shut down the gas turbine in the event that gas concentrations approach explosive levels.

The HSE dilution ventilation criterion provides the foundation for developing a design practice that inherently provides three levels of protection for managing hazardous gas leaks. Gas pocket volumes that remain below the dilution ventilation threshold are adequately diluted and gas turbine units can safely continue to operate under these conditions. When gas pocket volumes exceed this threshold, the leak is presumed under control but an alarm must sound giving the operator a warning to safely shut down the gas turbine. A high level trip provides the third level of protection, which assumes the leaks are out of control and will automatically shut down the gas turbine.

Our design practice is to place an array of three gas detectors in the ventilation extraction duct to ensure redundancy in the detection system. Appropriate location of the gas detectors and adjustment of their set points can provide an efficient warning of gas accumulation within the compartment, given a suitable criterion against which to measure dilution ventilation. Utilising CFD modelling, a direct relationship can be established between the %LEL of the detector setting and the size of the gas pocket for the worst case leak event. This information is then used to set the thresholds for safe operation of the gas turbine.

Our approach employs two threshold limits on gas accumulation. These are based on the 50%LEL and 100%LEL gas pocket sizes, i.e. the volume of the gas cloud that exists within the enclosure at concentrations in excess of 50% and 100%LEL respectively. Note that for any given leak rate the 100%LEL gas pocket is typically considerably smaller (e.g. factor of 10 or more) than the 50%LEL gas pocket.

The first limit is the HSE dilution ventilation limit, i.e. the point at which the size of the 50%LEL gas pocket exceeds 0.1% of the enclosure net volume. This typically occurs for a concentration of less than 10%LEL at the gas detectors. Any single detector reading this level will cause a "high" alarm. A typical example was shown in Figure 2. Note that this limit contains a built-in factor of safety since the concentration used to define the gas pocket size is a factor of two lower than the lower explosive limit.

The second limit marks the point at which the gas pocket hazard has escalated, such that the 100%LEL gas pocket size approaches 0.1% of the enclosure net volume. A safety margin is included by adopting a conservative value for the detectors' set point to ensure that the gas pocket is kept well below the 0.1% criterion value. This second limit is typically marked by concentrations in the range of 10% to 25%LEL at the gas detectors. If two of the three gas detectors read at this level, then a "high-high" alarm is initiated and the gas turbine will automatically trip.

In order to prevent a system being designed that generates alarms for extremely low or for inconsequential leak rate events, a calculation is also made of the type of leak event that would cause a detector alarm, using the approach laid out in Reference 3. This should typically be a minor flange leak, since other leak events are either too small (and will be promptly diluted) or too rare (the system would never alarm). If such a situation exists, then further design work is typically required.

Finally, our design practice also considers the effects of fuel ingestion into ventilated enclosures from external leaks, and fuel emission from enclosure vents. The former can lead to hazardous gas build-up within the enclosure, which is likely to contain an ignition source. The latter can result in the emitted fuel passing over an external ignition source, with consequent combustion. Utilising the tool previously discussed in section 3.3, steps are taken to guard against these situations by identifying boundaries around the enclosures, and by preventing the location of potential leak sources (e.g. piping flanges) and ignition sources within these boundaries.

5 CONCLUSIONS

WS Atkins has worked with GE Power Systems to develop protection systems for their gas turbine plants that employ dilution ventilation as a basis for safety against accidental gas leaks. This approach relies on an integrated ventilation and gas detection system, designed with the aid of computer simulations and field testing. The forthcoming ATEX Directives will require such an approach to be taken by all manufacturers wishing to market equipment within the EU. This paper has provided a description of the evolution of the relevant safety standards within the UK, and has shown how GEPS has developed its design processes in the light of recent findings. Brief discussions of the effect of liquid fuel leaks and hazardous fuel ingress have also been given.

6 REFERENCES

(1) Santon,R.C. "Explosion hazards at gas turbine driven power plants." *ASME 98-GT-215.*

(2) International Standard IEC 79-10, Third Edition, 1995 – 12, Electrical Apparatus for Explosive Gas Atmospheres, Part 10, Classification of Hazardous Areas.

(3) Cox,A.W., Lees,F.P. & Ang,M.L. "Classification of Hazardous Locations." *Institution of Chemical Engineers, 1990.*

(4) "Health and Safety at CCGT and CHP Plant." *HSE Interim Advice Note, 1997.*

(5) "Control of safety risks at gas turbines used for power generation." *UK HSE Guidance Note PM84, 2000.*

(6) European Commission "ATEX Directives and their application to Gas Turbines," *http://europa.eu.int/comm/enterprise/atex/gasturbines.htm, 2001.*

(7) Gilham, S., Cowan, I.R. & Kaufman, E.S. "Improving gas turbine power plant safety: the application of CFD to gas leaks." *Proc. Instn. Mech. Engrs., Vol. 213 Part A, pp475—489, 1999.*

(8) Bowen,P.J. & Shirvill,L.C. "Combustion hazards posed by the pressurized atomisation of high-flashpoint liquids." *J. Loss Prev. Process Ind., Vol. 7 No 3, pp233—241, 1994.*

C603/022/2003

MagMax™ – the development of a revolutionary sealless pump

K BLACK
HMD/Kontro Sealless Pumps Limited, Eastborne, UK

SYNOPSIS

MagMax™ is a revolutionary new product, drawing on the best features of Mechanical Sealed Pumps (MSP), Canned Motor Pumps (CMP) and Magnet Drive Pumps (MDP) to deliver a totally unique product that is set to change the pumping market.

The MagMax™ Canned Magnet Drive Pump (CMDP) was launched to pump specifiers and users Worldwide in June 2001, since then it has received critical acclaim for its combination of simplicity of design (and hence low cost of ownership) of a Magnet Drive Pump and true Secondary Containment of a Canned Motor Pump.

This paper presents the technical challenges that were encountered during the development of MagMax™ and discusses the ways in which solutions to these challenges were developed. In particular, this paper discusses the following:

- The development of a unique high efficiency motor (greater than 10% more efficient than typical canned motor pumps).
- The development of the 'Vapour Detector' that, uniquely, detects the onset of dry running before pump failure has time to occur.
- How the design concept allows for the use of Silicon Carbide bearings, as standard, without the necessity for any special bearing alignment techniques.
- Achievement of certification for use in potentially explosive atmospheres.
- Development of a product that utilises components and manufacturing processes from HMD/Kontro's existing magnet drive pump – hence reducing manufacturing cost, lead-time and spare parts inventory.

The paper then concludes by discussing the benefits that MagMax™ brings to the user.

1. INTRODUCTION

Increased environmental awareness, demand for the highest reliability coupled with ease of maintenance, and falling prices have combined to drive the adoption of sealless technology in more diverse application areas, where traditionally MSP would have been used.

Figure 1. Shows an application pyramid indicating where differing pumping technologies are used. Across the base of the pyramid is the size of market, whilst the severity of service and equipment price increases as the pyramid is climbed.

Figure 1. Application pyramid for pumping technologies

The position of the break between sealless and sealed technologies is predominantly driven by a cost versus risk analysis. The main reason why MSP have greatest market share is that they are perceived as lower cost than sealless pumps. This perception is not always correct and enlightened users who look at through life costs have identified that sealless pumps are a lower cost solution than sealed pumps on many applications.

As a pump manufacturer looking for growth, clearly a manufacturer of sealless pumps has the greatest opportunity by moving down the pyramid where the market opportunity rapidly expands. The CMP technology at the top of the pyramid has the biggest market opportunity available.

HMD Sealless Pumps Ltd. (HMD), a leading supplier of MDP believed that they could increase their business by opening up opportunities in different segments of the application pyramid. As both CMP and MSP are mature technologies with established and respected manufacturers, it would be difficult to break into these markets without a unique selling proposition (USP). In other words a "me too" product would not be successful. A market survey conducted with key customers about their' pumping needs and what they would like from a new pump technology resulted in the following feedback:

- We've lost our rotating experience – simple is best
- We get paid to be on-line – the equipment must be reliable
- Equipment must withstand upsets – it must be robust
- We would like one supplier for all technologies, or better still one technology for all applications
- Wire to water efficiency must be better than existing technologies
- The new product should have the best features of the three existing technologies

The last message created a stir within the MDP business as for years CMP and MSP were seen as inferior technologies. Could the two other technologies offer benefits to the customer over MDP? Clearly the answer was yes, borne out by the installed base. Therefore another customer survey was carried out on what they liked about each technology. The results are shown in Figures 2 – 4.

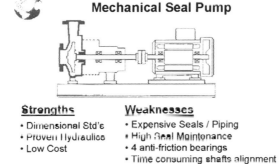

Mechanical Seal Pump

Strengths
- Dimensional Std's
- Proven Hydraulics
- Low Cost

Weaknesses
- Expensive Seals / Piping
- High Seal Maintenance
- 4 anti-friction bearings
- Time consuming shafts alignment
- Costly baseplate

Figure 2. Strengths and weaknesses of MSP

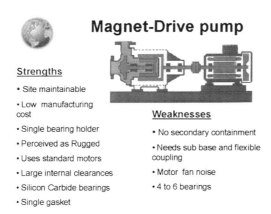

Magnet-Drive pump

Strengths
- Site maintainable
- Low manufacturing cost
- Single bearing holder
- Perceived as Rugged
- Uses standard motors
- Large internal clearances
- Silicon Carbide bearings
- Single gasket

Weaknesses
- No secondary containment
- Needs sub base and flexible coupling
- Motor fan noise
- 4 to 6 bearings

Figure 3. Strengths and weaknesses of MDP

Canned motor pump

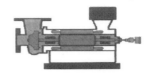

Strengths	Weaknesses
• Secondary containment	Factory repair
• Low Noise	Small running Clearances
• No Sub Base	Expensive to manufacture
• Simple design	Perceived as fragile
• 2 bearings	Generally Carbon Bearings
	Two Gaskets

Figure 4. Strengths and weaknesses of CMP

With the above information, a Sales and Marketing specification was written for a new product called MagMax™ which had the features shown in Figure 5

MagMax - combined to win

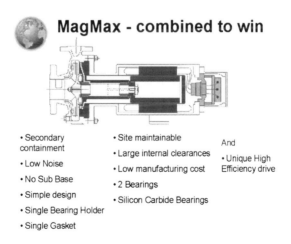

• Secondary containment	• Site maintainable	And
• Low Noise	• Large internal clearances	• Unique High Efficiency drive
• No Sub Base	• Low manufacturing cost	
• Simple design	• 2 Bearings	
• Single Bearing Holder	• Silicon Carbide Bearings	
• Single Gasket		

Figure 5. Required features of new product

To maximise the opportunity for producing MagMax™ at a low cost, the decision was taken to utilise as many components from the existing MDP range. This ensured that economies of scale would be realised which would also have a knock-on affect to the manufacturing costs of existing MDP.

2.   DESIGN

The Sales and Marketing specification for MagMax^{TM} was a shopping list of would likes, must haves and don't even think about showing us something without these included. These wishes were prioritised and several design alternatives put forward which satisfied these needs. For each design a technology review and gap analysis was carried out based on existing knowledge and capability within the business and a risk analysis versus return carried out. The key factors which drove the chosen design was the option which included the development of a unique high efficiency drive and also being totally site maintainable.

The design considerations were broken down into six elements to identify what development activities were necessary:

*   Hydraulics
*   Bearings
*   High efficiency drive
*   Maintainability
*   Electrical connections
*   Hazardous area certification

2.1   Hydraulics

As efficiency is important, all MDP hydraulics being used on MagMax^{TM} were reviewed and two new hydraulic sizes were designed and tested. The new hydraulics designs were proven to be 13 to 20% more efficient than the existing designs. This is not covered in any more detail in this paper, as hydraulic design principles are not unique to this product.

2.2   Bearings

Proven silicon carbide bearings that are standard on the MDP were selected to give interchangeability across product ranges and low manufacturing cost. Silicon carbide bearings are used due to their excellent chemical compatibility, hardness, heat transfer and load carrying capability. Although proven silicon carbide bearings were selected, a development program was required to ensure reliability as discussed later. Carbon bearings were not considered as they may wear and regular maintenance would be required.

2.3   High efficiency drive

HMD's core competency is magnet coupling design and manufacture. However the magnet couplings used on MDP could not be used because secondary containment is required from MagMax^{TM}. MDP have a rotating magnet ring driven by an external source, which cannot be sealed to give true secondary containment. CMP have secondary containment as standard because the rotating magnet ring is replaced by a static motor stator, which is contained in a pressure vessel. The obvious solution to satisfy the secondary containment requirement is to use a static motor stator the same in principal to the CMP. However on CMP the rotor which is driven by the rotating magnet field in the stator, is similar to an induction motor rotor but with a much bigger air gap. This results in a very inefficient drive with high slip (difference in speed of rotor compared with speed of rotating field in stator).

Looking at various options, it was decided to develop a range of permanent magnet motors for use in the MagMax^{TM} range due to their high efficiency capability and due to the magnet coupling expertise at HMD.

Permanent magnet motors work by having a static motor stator with a rotating magnet field driving a rotor containing permanent magnets. The rotor runs synchronously with the rotating stator field. The design of the stator winding configuration, the magnet configuration in the rotor and an induction device for getting the rotor up to speed were all key development activities which are discussed later.

2.4   Maintainability

Although MagMax™ must be designed to be as reliable as possible, if a breakdown occurs complete field maintainability must be possible. The MagMax™ drive uses a static motor stator similar to that used in CMP. However, CMP have non field replaceable stators which results in expensive and time-consuming return to factory repairs. A key development activity was to assess the thermal impact caused by a replaceable stator concept to ensure that this was a viable option.

2.5   Electrical connections

Passing current from the external mains supply to the internal motor stator through a secondary containment vessel is quite difficult. Existing CMP use complicated and expensive terminal blocks and feed throughs. The CMP solution does not allow for site maintainability so this approach could not be carried through onto MagMax™. The development activity was to test suitable proprietary (low cost, proven and reliable) feed throughs for pressure retention, thermal performance and ease of assembly. As proprietary low cost feed throughs meeting the above criteria were sourced and proven suitable through testing, this subject is not covered in detail in this paper. However it should be noted that significant manufacturing engineering input was required to ensure that during welding, temperatures were kept below a predefined level.

2.6   Hazardous area certification

To maximise the opportunity for Sales to offer MagMax™ into the largest number of applications, the new product must be suitable for use in potentially explosive atmospheres as well as in safe areas. A development program was put in place to provide an acceptable thermal and hardware design to enable certification to EExde IIB + H_2 T4 which would meet the requirement for the majority of potentially explosive atmospheres.

3.   RELIABILITY

Pump users get paid to be online. They do not get paid to repair pumps. Downtime costs money. However it is packaged, users want to purchase, install and commission a pump and then ideally forget about it, although it is reluctantly accepted that a certain level of maintenance is required. However, this maintenance should be planned and not as a result of failure thus minimising the impact on production.

When designing MagMax™ and the driver to use existing parts from the MDP range, the temptation was to duplicate many existing features. However it is a business philosophy that new products should be ten times more reliable than old products. Therefore a Marciano chart was produced based around the current MDP to identify any weaknesses which ultimately impacted on reliability. Identified weaknesses were designed out. Where new component parts were generated, proven manufacturing methods were used to enable existing cellular manufacturing and common set-ups to be used to drive down costs.

A report by a leading chemical petrochemical company showed that seals and bearings cause 80% of pump failures – see Figure 6. If these can be designed out then increased reliability should follow.

Why do Pumps Fail?

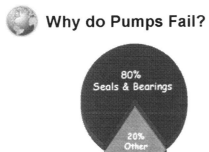

Figure 6. Why do pumps fail?

MagMax™ has only one primary containment seal which is a static gasket and this is the minimum possible for a maintainable product. MagMax™ has two bearings, which is the minimum number practical for this type and size of equipment. These product-lubricated bearings are identical to those proven on the MDP range. External bearings have been designed out

The use of non-wear silicon carbide bearings is the ultimate in "fit and forget" technology and has the potential to give the highest level of reliability. The word potential is used because silicon carbide has been used in pumps for some years but pumps still fail. The reason for this is that users run pumps dry and silicon carbide fails rapidly under boundary conditions. Pump suppliers pass the blame onto the user for the failure which, whilst this is true, the user will often not be able to change his process without major cost and will look for a pump that can run dry. It was therefore necessary to build into the design of the MagMax™ pump the ability to cope with dry running without the pump failing. Various attempts have been made by pump suppliers at producing a dry running bearing design such as diamond coated or carbon impregnated silicon carbide, but these have limited capability and do not provide a solution. Pump manufacturers whilst still tasked with searching for the Holy Grail of the dry running bearing, must until this becomes available offer a solution to the dry running problem if highest reliability silicon carbide bearings are to be used.

Electric motors driving pumps have reliability issues due to bearings and insulation breakdown caused by moisture or over-temperature. External rotating parts have been designed out. Insulation selection and over-temperature protection was required as a development activity.

4. DEVELOPMENT ACTIVITIES

The following sections outline a brief summary of some of the important development activities. Not all issues are covered due to their' sensitivity and intellectual property rights, however this paper gives enough insight for the reader to appreciate the process and resulting benefit to the user.

4.1 Silicon carbide bearing reliability

The ability for MagMax™ to cope with dry running was debated and whilst good results were obtained with silicon carbide under certain conditions of load and speed it was considered too complicated a subject to cover all conditions. Therefore the stance was taken that the pump will not be sold as a pump with dry running capability. However, the design would have the ability to detect the presence of fluid and if the pump started to run dry or vaporisation of fluid occurred, the pump could be automatically shut down before bearing damage could occur. To meet this need, HMD developed a device called a "Vapour Detector" which is welded in the containment shell and sits in the pumped fluid – see Figure 7 below.

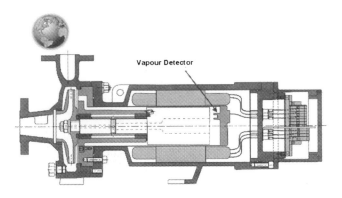

Figure 7. Location of Vapour Detector

The vapour detector is basically a tuning fork which vibrates at its natural frequency in air and is damped when surrounded by any fluid. When damped (liquid present), the signal to the control box keeps a relay closed, and when the pump dry runs or the liquid vapourises the probe will vibrate at it's natural frequency and the relay will open. This relay if wired into the motor circuit will cause the pump to be stopped as soon as dry running is detected. To avoid phantom trips caused by slugs/pockets of air, the control box which is mounted in the terminal box has an adjustable timer. The vapour detector is a standard option on MagMax™ and is suitable for use with many voltages making it a universal solution to detect the onset of dry running.

Figure 8. Vapour detector in end plate

Figure 9. Containment shell fitted with vapour detector

The use of a clear acrylic containment shell with the pump being driven from the suction enabled air injection tests to be carried out to check the operation and ideal position for the vapour detector. This test rig shown in Figure 10 was also used to ensure that after shutdown due to dry running and then re-establishing wet suction, the pump would re-prime. If the pump was not self-venting and was air locked then the vapour detector would not allow the pump to be restarted. Although the vapour detector is working correctly and saves the pump from damage, the process cannot be run although it is in an acceptable condition. Testing showed on some prototype models that air locking did occur and the design was modified to incorporate a small vent orifice back to suction.

Figure 10. Flow visualisation and vapour detector test rig

Silicon carbide bearings must be well aligned to eliminate point contact. Traditional CMP use carbon bearings (which wear) because the nature of the design makes it difficult to get the required level of alignment needed for silicon carbide. Some CMP use silicon carbide bearings but they are complicated and expensive assemblies. To ensure repeatably good alignment, the proven designs used on MDP were used on MagMax™ which incorporate a one piece bush holder – see Figure 11. Being one piece the bearing locations can be machined together guaranteeing alignment.

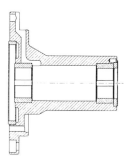

Figure 11. One piece bush holder enables silicon carbide bearings to be used

This philosophy is proven on the MDP ranges and has been carried across to MagMax™.

4.2   High efficiency drive

Standard induction motor stators are constructed from steel laminations with cut-outs (slots) in the bore into which are wound numerous copper wires. The connection of a potential difference (voltage) across these wires results in current flowing and a resulting magnetic field. When current flows there is a resultant temperature rise in the wires which is proportional to the square of the current.

These copper wires are tightly packed and in close proximity. As they are carrying current they need to be insulated to ensure that they do not "short-out". Various insulating materials and methods can be used and the selection of the correct method will be determined by the maximum temperature expected by the design. The temperature of the windings will be governed by three factors :

- Current drawn
- Presence of any external heating
- Degree of cooling

The combination of the above factors results in a winding temperature that the insulation must be capable of withstanding to avoid breakdown. Insulation systems are designated by class with maximum operating temperature as follows :

Table 1. Insulation classes versus maximum allowable temperature

Insulation Class	Ambient temperature °C	Temperature rise °C	Maximum temperature °C
A	40	60	100
E	40	75	115
B	40	80	120
F	40	105	145
H	40	125	165

Higher temperature classes of insulation exist but are expensive and not usually commercially available. MagMax™ has class H insulation as standard to provide a wide service envelope at rated output powers.

The temperature in the windings can be controlled by varying the three factors. For example, increasing the process temperature can be offset by derating the output load and hence current. The affect of load and process temperature on the winding temperatures and other parts of the pump can be thermally modelled with appropriate software or by actual testing. This is covered in more detail later.

Standard induction motors are force air cooled by an auxillary fan mounted on the non drive end of the rotor shaft.

The stators on CMP and MagMax™ are contained within an enclosed housing and therefore it is impossible to force air cool except by a separate external fan. However on both CMP and MagMax™ there is a flow of liquid (pumpage) through the gap between the stator and rotor which will impact on the temperature of the windings. This flow of liquid is taken from the main pumpage and is therefore at the same temperature as the process. In the case of MagMax™ it is possible to have a process fluid temperature of 120°C.

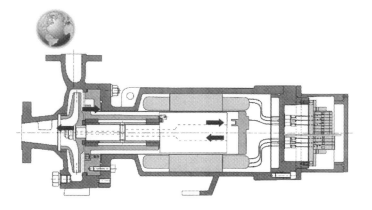

Figure 12. Cooling flow through MagMax™.

The ability of the design to dissipate heat generated in the windings is influenced by the degree of contact of the stator within it's housing and with the stator can. Any air gaps will affectively act as an insulator. As MagMax™ has a replaceable stator concept, by design there are air gaps present between the stator and housing and the stator and stator can. There is also a large air gap around the stator end windings. To minimise the air gaps and produce a good thermal design the following features were needed :

- Encapsulation of end windings to fill air gap
- Heat transfer media to fill small air gaps on outside and inside of stator.

To protect the windings from excessive temperature due to process upset conditions, over-temperature protection is provided. Two sets of triplet thermistors are embedded in the end windings with set points of 160°C and 180°C. One can be configured as an alarm and the other as a trip. These thermistors will protect the insulation from seeing too high a temperature and breaking down. The life of the insulation and hence motor life is related to temperature. Increase in temperature over class rating severely affects life. The use of the thermistors will ensure reliability of the winding insulation.

4.2.1 End winding encapsulation
To produce a specification for the encapsulation, not only were the thermal requirements considered but also through life considerations including reliability. This resulted in the following wish list of encapsulation characteristics :

- Electrical isolation of stator windings from containment shell
- Increased surface area and gap filling capability to assist in heat dissipation from stator end windings

  C603/022/2003 © HMD/Kontro Sealless Pumps Limited 2003

- Mechanical protection of stators during storage and assembly
- Prevents moisture from entering windings and causing insulation breakdown
- Provides a level of protection against chemical attack

From the above characteristics the encapsulation material should have the following properties:

- suitable for continuous use at 180°C without plasticizing or breaking down
- Will not absorb moisture
- electrical insulator
- resistant to temperature cycling
- dimensionally stable
- good thermal conductivity
- mechanically rigid

Working with a reputable motor manufacturer, a two part epoxy resin was trialed and proven successful. Part of the trial was to assess the thermal profile with and without encapsulation.

Figure 13. Encapsulation of stator

Figure 14. Hot oil test rig for checking thermal stability

The encapsulation resulted in a 50°C temperature reduction which meant that the motor could be used with a higher process fluid or give a higher output power for a given process temperature. The encapsulation was also thermally checked for stability and expansion on a hot oil test rig pumping oil at maximum process temperature of 120°C.

4.2.2 Stator winding / rotor magnet configuration

MagMax™ is driven by a permanent magnet motor. Permanent magnet motors work by having a static motor stator with a rotating magnet field driving a rotor containing permanent magnets. The motor stator has a magnet field rotating at :

Speed = frequency of mains supply (Hz) x 60 / number of pole pairs
Therefore a 50 Hz supply for a two pole motor rotates at 3000 rpm and 3600 rpm for 60 Hz.

The rotor runs synchronously with the rotating stator field and therefore the rotor rotates at a constant speed regardless of output load. However if the rotor output load is taken well beyond it's rated load, the rotor will desynchronise with the rotating stator field. The desynchronisation affect is similar to synchronous magnet drive pumps, however the MagMax™ drive has a unique advantage. Magnet drive synchronous pumps when desynchronised will not re-sychronise unless the pump is stopped. The desynchronised inner magnet ring will run very slowly and provide inadequate cooling and the pump will rapidly overheat.

Uniquely, MagMax™ can be resynchronised whilst running by backing off the load to bring it back into it's rated envelope. This ability is due to the patented design of the MagMax™

rotor. If the load is not backed off, the MagMax™ rotor will run at a reduced speed but adequate enough to provide cooling flow until the increased motor current thermally trips the pump. This is a unique feature and provides a level of robustness under off design conditions.

Induction motors used to drive MSP, MDP and built into CMP have a rotating stator field in concept the same as MagMax™, however their rotor design has no magnets. The magnetic attraction of the rotor to the stator field is caused by currents induced in conductors in the rotor which in turn produces a magnetic field. The induced (hence name induction motor) magnetic field is not synchronous and the rotor runs at a reduced speed. The higher the rotor output load the lower the speed. In pumps the rotor is connected to the impeller and hence reduced speed will give reduced output. MagMax™ with a true synchronous motor runs faster than all of the other technologies and hence gives a greater hydraulic performance (approximately 7% more flow and 15% more head at 50 Hz) within an ISO or ANSI envelope. This means that more can be obtained from a given size and in some cases a smaller, lower cost liquid end can be selected.

The design of permanent magnet motors provides the following development challenges :
- Starting capability
- Selection of magnet material for torque rating
- Selection of magnet material to resist demagnetisation

When a mains supply is connected across the permanent magnet motor stator, instantaneously the magnet field is rotating at full speed. The rotor and connected parts such as impeller have an inertial mass which is too large to allow the rotor magnets to remain synchronised for more than an instant. Therefore the rotor will never get up to speed. There are several ways of resolving this problem as follows :

- Use an inverter
- Slave drive by another device until rotor is up to speed
- Design an induction capability into the rotor

The first two options add expense, complication and more parts to go wrong and therefore do not fit into the brief of MagMax™ being simple, reliable and competitively priced. Therefore a unique rotor was developed which contains permanent magnets and also an induction device. The theory was that the induction device worked the same as in an induction motor and ran the rotor from stationary to a speed where the permanent magnets locked in and took over. Once running synchronously the induction device has no function. The design of the induction device needed to be low cost, easy to build into the rotor and compact to allow room for the magnets. If the induction device took up too much room, the magnet volume left would limit the torque and hence output capability of the MagMax™ motor. To reduce the development timescale and hardware costs, computer software was used to model the starting capability of various induction devices to check whether they would allow synchronisation. Typical results are shown in Figure 15 which indicate that only induction device 5 will synchronise.

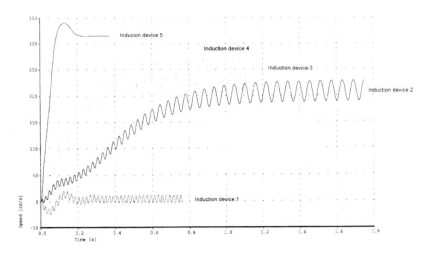

Figure 15. Results of rotor synchronisation analysis

Having designed a suitable induction device, various magnet materials were assessed with a view to ensuring maximum torque versus magnet volume, lowest purchasing cost and ease of rotor build. The standard magnets used on the MDP range were firstly assessed for suitability, however other materials and shapes were considered to maximise output and efficiency. The use of lower cost shaped magnets seemed attractive and would increase magnet volume and hence torque, but these had BH curves at the higher operating temperatures with a "knee" in the second quadrant – see Figure 16.

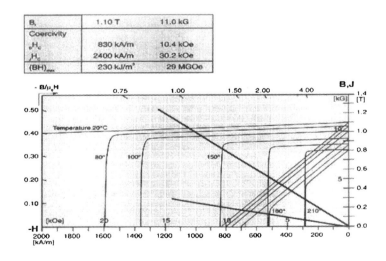

Figure 16. Magnet properties – Flux versus coercivity (BH) curves

    C603/022/2003 © HMD/Kontro Sealless Pumps Limited 2003

For example, if the magnets were subject to 180°C and the working point of the magnets moves down the BH curve from the upper line to beyond the knee (lower of the two lines), the magnets will permanently loose strength and the torque rating reduce. Classic theory can be used to calculate the position of the load line and hence magnet material selection. However this led to some painful experiences as the classic theory does not take account of demagnetisation at the flanks of the magnets. The use of computer modelling software enabled a more accurate analysis to be carried out and showed localised demagnetisation confirming actual test results. Figure 17 shows the finite element result. The outer section with the slots represents the stator and the inner section the rotor. The various shadings indicate flux density. A good design would have a flux density of approximately 1.0T in the magnets and a maximum of 1.4T in the stator core and rotor backing iron. In the model several problems are highlighted leading to performance problems confirmed during testing. Firstly the magnets show heavily shaded for a third of their arc. This area is at a flux density of less than 0.2T and therefore at this area the load line is below the knee (see Figure 16) on the BH curve and permanent strength loss is occurring. Secondly areas of the stator are showing saturation. Although not ideal, the degree of saturation is small and is not a cause for concern. However the current will be slightly higher than ideal.

The use of the software to model the impact of changing the magnet arc angle, magnet thickness and magnet material resulted in improved output, lower current and hence lower temperature rise and a higher power factor. The biggest gain was an improvement in efficiency. Changing the magnet material and shape resulted in a 23% reduction in current at full load and a resulting increase in the power factor, a reduction in temperature in the windings by 20°C and an efficiency improvement of 4%. All key motor performance parameters.

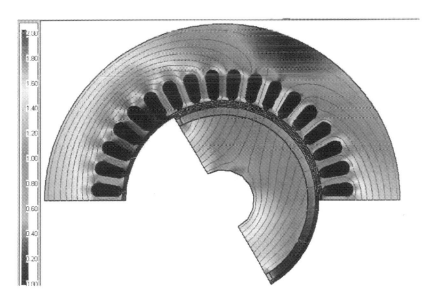

Figure 17. Results of finite element analysis of motor and rotor design

Another benefit of using a permanent magnet motor design in MagMax™ is that the starting current is only approximately 400% of full load current compared to approximately 600-700% of full load current on induction motors. This means that the size of the starters and associated equipment can be reduced and costs saved.

4.2.3 *Wire to impeller shaft efficiency*

Wire to water losses for pumps can be defined as the difference in power supplied by the mains to that available at the impeller shaft. This difference can then be used to calculate the wire to water efficiency. Typical losses are as follows :

- Motor fan
- Motor rolling element bearings – friction losses can be high if sealed bearings used
- Motor shaft seal - if used
- Winding – caused by I^2R copper losses
- Pump rolling element bearing - friction losses can be high if sealed bearings are used
- Can losses – metallic cans have hysteresis losses
- Rotor dynamic – caused by surface drag of rotating parts in the pumpage
- Rotor can losses – hysteresis losses in rotor can caused by difference in speed (slip)
- Product lubricated bearings – friction losses
- Seal and support system – caused by friction on seal faces and running the buffer support system if double mechanical seal

Table 2 summarises which technologies have the above losses.

Table 2. Losses versus pumping technologies

Loss	MSP	MDP	CMP	MagMax™
Motor fan	Yes	Yes	No	No
Motor rolling element bearing	Yes	Yes	No	No
Motor shaft seal	Yes	Yes	No	No
Winding (copper)	Yes	Yes	Yes	Yes
Pump rolling element bearings	Yes	Yes	No	No
Can	No	Yes	Yes	Yes
Rotor dynamic	No	Yes	Yes	Yes
Rotor can losses	No	No	Yes	No
Product lubricated bearings	No	Yes	Yes	Yes
Seal and support system	Yes	No	No	No

Published data for induction motors give efficiencies at various loads. Typically for high efficiency F1 motors they are about 85% efficient for smaller motors (5.5 Kw) and increase to approximate 90% on larger motors (22 Kw). The smaller motors are less efficient because the fan, bearings and seal losses are more significant and affect the efficiency to a greater extent.

For MSP, seal and support system losses are stated as small numbers but can be much higher depending on whether the seal is balanced and how the support system is maintained. Thermally circulated systems are more efficient than systems which have a circulating pump. Based on published data, test results and assumptions, the following is a comparison of wire to water efficiencies for the various technologies.

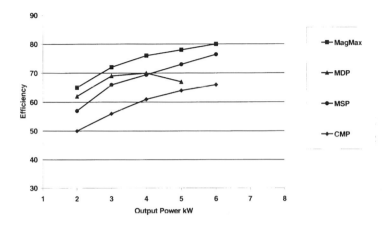

Figure 18. Wire to impeller shaft efficiency comparison

MagMax™ comes out with the highest efficiency due to the low losses in the stator and can. It should also be noted that the losses for MagMax™ are at 3000 rpm and not the reduced speeds of the other technologies. Therefore MagMax™ is more efficient and at the same time is delivering more performance.

5.   HAZARDOUS AREA CERTIFICATION

Equipment for use in potentially explosive atmospheres must meet certain design criteria to ensure that they cannot generate an ignition source to the surrounding atmosphere. Any explosion created within the equipment must be fully contained. Equipment for use in Europe must be designed in accordance with a set of standards EN 50014-19 dependent upon the protection philosophy to be employed.

Protection can be by various methods but the most common are :

d – explosion proof. Under normal conditions an ignition source is possible and therefore the housing needs to be designed to contain an explosion. Criteria are set for the design of flame paths which are dependent on gas volume and gas type within the equipment.
e– increased safety. Under normal conditions an ignition source is not possible and therefore the housing does not need to be designed to contain an explosion. Criteria are set for creepage and clearances between conductors to avoid tracking and arcing. These criteria are dependent upon voltage.

It is normal for users to specify that equipment must be certified by an approved body and be safety marked before acceptance into the hazardous area. The marking must meet the CENELEC requirements and constitutes the design philosophy of the equipment and where it is safe to be installed. The required design for MagMax™ is EExde IIB + H₂ T4. This means that :

**EExde = Explosion proof pump with increased safety terminal box
IIB+ H₂ = Suitable for areas containing IIB gas groups and also hydrogen
T4 = suitable for areas containing gases with ignition temperatures above T4 (135°C)

** MagMax™ has been designed and certified to be both EExde and EExd. EExd allows for the situation when the vapour detector control box is fitted into the terminal box.

Certification is carried out by a UKAS approved company and for MagMax™, Intertek Testing Services (ITS) were used. The certification process involved five stages :

- Approval of documentation and quality system to support manufacture of product
- Production of certified product drawings detailing design features and compliance with standards
- Thermal performance testing to prove compliance with maximum surface temperature class
- Flame transmission tests to prove that an explosion cannot escape to the surrounding atmosphere
- Training of personnel

The first was relatively straight forward as HMD have had an approved ISO compliant quality system for many years. Some procedures needed modifying to cover certified products.

Once the design was finalised, the certified drawing was relatively easy to produce with the help of ITS. A complication was that key features and dimensions that ensure that the CENELEC standards were met would need highlighting within the business. This was needed to ensure that they could not be changed or out of specification parts accepted without approval against the standards. To achieve this, all drawings affecting the certified product were identified with a special symbol and note and critical dimensions flagged. CAD files had these symbols and notes embedded within them. A procedure was written to control this activity.

The thermal testing was a continuation of previous thermal testing but with thermocouples on other critical areas such as insulation, cable glands, etc in the increased safety terminal box. Either through testing or modelling, data must be available to show that the T class must not be exceeded and also that no component reaches a temperature which is beyond it's allowable temperature. The equipment was run at rated output at maximum process temperature until the temperatures stabilised.

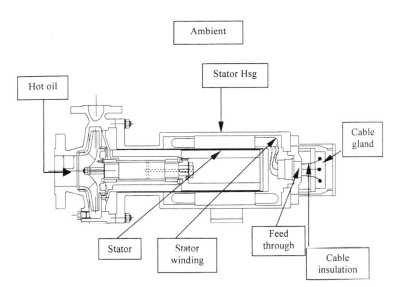

Figure 19. Location of thermocouples during thermal analysis

Running in parallel with the thermal work was flame transmission testing. A MagMax™ pump was modified to include gas entry and exit ports, a piezoelectric pressure transducer and a spark plug. A gas mixture of 31% ±1% hydrogen in air was allowed to purge the stator housing for sufficient time for the correct mixture to be obtained throughout. The gas mixture was ignited using a low energy, high voltage discharge and the resulting pressure recorded. This test was repeated at an ambient temperature of -40°C. The stator housing was then subject to a pressure test to ensure that no distortion or visible damage occurred. This is an arduous test for the stator can as cylinders subject to external pressure can easily buckle. MagMax™ has been designed to withstand this buckling pressure. The parts of the pump designed as flame paths were partially separated to meet the test standards, the pump was then placed in a plastic sheet walled explosion vessel and filled with 28% hydrogen in air. The plastic vessel was also filled with the same gas. The gas within the pump was then ignited and if the pump failed the flame transmission test then the external gas blanket would be ignited. The gas was not ignited and MagMax™ passed the flame transmission test.

Traditional pump assembly at HMD involves mechanical engineering and very little electrical engineering. As MagMax™ has a built in motor with associated wiring, new skills had to be learnt throughout the business. Electrical theory and practical courses were organised by an outside training body resulting in twenty employees passing the course. Only these employees are authorised to work on MagMax™ pumps. Special build and test procedures are in place and final sign off prior to shipment has to be authorised by one of two certified product engineers.

5. MAGMAX™ – THE BENEFIT TO THE USER

Through the design and development process the original brief was constantly revisited and

reviewed to ensure that the final product would meet the customer's wishes :
- We've lost our rotating experience – simple is best
- We get paid to be on-line – the equipment must be reliable
- Equipment must withstand upsets – it must be robust
- We would like one supplier for all technologies, or better still one technology for all applications
- Wire to water efficiency must be better than existing technologies
- The new product should have the best features of the three existing technologies

MagMax™ has been designed and developed to be a product that sits at the top of the application pyramid in terms of safety but drives down the application pyramid by nature of it's features and more importantly cost. Users now have the ability to select one product to cost effectively meet most applications.

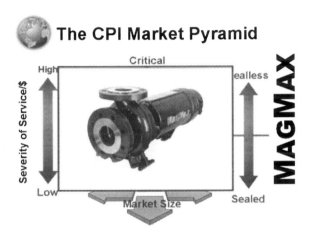

The CPI Market Pyramid

Figure 20. MagMax™ the pumping solution for all areas of the pyramid

Although in some cases MagMax™ maybe a higher capital cost than other technologies, by the time installation and maintenance costs are included as well as operating costs, MagMax™ has a lower through life cost than any other technology. Although costs are not covered in this paper, comparisons are available if required from the author's company.

MagMax™ is the highest reliability technology as there are the minimum practical number of seals and bearings, non wearing silicon carbide bearings, dry running detection and thermal protection of the stator. The key to reliability is to ensure that MagMax™ is not allowed to run dry. The installation of the vapour detector will protect against this. The vapour detector provided the best technology available compared to other methods used by pump suppliers.

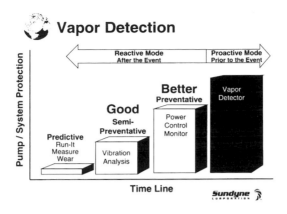

Figure 21. The MagMax™ Vapour Detector ensures bearing reliability

A unique rotor construction developed for MagMax™ provides robustness in case of running under off design conditions which causes the rotor to desynchronise. The design allows enough time for thermal protection to pick up the condition and shut the pump down before any damage can occur. Unique to synchronous couplings, MagMax™ gives the ability to resynchronise without shutting the pump down.

The unique high efficiency drive is capable of direct online starting without using any ancillary devices and allows the reduction in starter sizes due to reduced starting current over induction machines. MagMax™ is totally field maintainable, a first for a product with true secondary containment.

In summary, the benefits to the user are :

- One type of pump for all areas of the application pyramid makes specification easy
- Reduced spares inventory opportunity if MagMax™ is used for all applications
- Lowest true life cycle cost
- Fully site maintainable
- Low noise level (less than 68 dB(A))
- Simple design
- Robust
- Unique high efficency drive reduces operating costs
- Reliability through design and protection
- Reduced starting current enables reduced switch gear size and costs
- Increased performance from same ISO or ANSI hydraulic size
- Secondary containment as standard
- High quality from proven MDP manufacturing processes
- Short lead times for pumps and spares
- Fit and forget design utilising silicon carbide bearings
- Compact design minimising valuable floor space
- Simple installation

- Interchangeability of parts with MDP range
- Unique rotor design with inbuilt protection if rotor desynchronises
- Direct on line starting with no assistance required
- Thermal protection built in as standard
- No emissions, costly clean-ups, damaged reputation
- Reduced insurance premiums
- Safe working environment
- Clean working environment
- Environmentally friendly product

The Development Team have succeeded in bringing to market a product that meets the customers' needs as specified in the Sales and Marketing specification. However, through lessons learned during the development program, there are opportunities to bring even greater benefits to the user. MagMax™ generation 2 is evolving and we believe that it will be even more revolutionary than the product presented in this paper.

C603/033/2003

Application of a diagnostic system for mechanical seals

C SCHMIDTHALS
Burgmann Dichtungswerke GmbH & Co KG, Wolfratshausen, Germany

ABSTRACT

The life cycle costs of a pump are determined by the consequential cost associated with pump failures. The failure of a pump will in many instances be caused by the failure of the mechanical seal. This is often caused by non permissible operating conditions. A system of reliably forecasting the remaining life time of a mechanical seal must thus be established in order to replace them during normal non productive periods. This diagnostic system was developed by a manufacturer of mechanical seals in collaboration with a pump manufacturer. The required diagnostic software controlling the actual life time computation comprises user interfaces and the so-called life cycle algorithm.

1 FAILURE RATES OF MECHANICAL SEALS

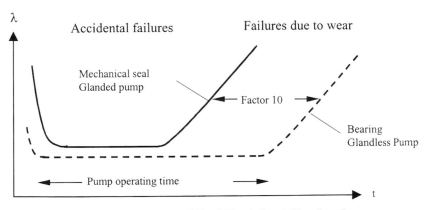

Figure 1. Typical Failures Rates λ(t) of Glanded and Glandless Pumps.

The failures of glanded pumps are to a great extent, as shown in Figure 1, due to mechanical wear at which the wear on mechanical seals begins earlier by the factor 10 than e.g. the wear on bearings of glandless pumps with hydro-dynamic lubrication (1). As shown in Figure 1 at the beginning of the life cycle the failure rate initially strongly decreases with increasing time of operation. This can be attributed to the failure of mechanical seals and pump parts due to wrong installation and/or manufacturing defects. At longer operating times the failure rate remains constant at a low level. In this period the failures are caused mainly by system-related non permissible operating conditions. At even longer operating times again an increase of the failure rate can be observed, that is related to the natural wear of mechanical parts in the pump and the mechanical seal. It must be observed that the failure rate λ is a statistical value depending on a variety of diversely possible operating conditions. This value can only reflect the probability of a failure but not its definite time. Two categories of causes of the failure of a mechanical seal can be distinguished:

- System-related causes, non permissible operating conditions
- Natural wear of the mechanical seal

From this introduction it becomes clear that a system for increasing safety must have the following properties:

- Detection of non permissible operating conditions which can be critical for the mechanical seal
- Detection of the natural wear of a mechanical seal and computation of its remaining life time

2   PRINCIPLES OF THE WEAR OF MECHANICAL SEALS

It is extremely difficult to make a generalized statement on the expected wear of mechanical seals. One of the reason for that is the multitude of influences on the wear as well as their mutual influences on to each other (2). In conjunction with the here introduced diagnostic system there are mainly the influences of pressure and temperature that are of interest with regard to the wear and thus the life time of a mechanical seal. By life time the time period is meant during which the length of the seal ring is reduced to a minimum due to natural wear and at which further wear would lead to a definite leakage.

After the theory of Archard (3) and Holm (4), the so-called adhesive wear is a linear function of the slide pressure.

This simplified relation can be used with caution only since the influences of the fluid on the wear, the frictional conditions and the temperature have not been taken into account. It must specially be observed that a higher slide pressure will generally lead to a higher temperature in the seal gap thus adding on to the influence of the temperature of the product.

Early investigations of the life expectancy of balanced mechanical seals (5) yielded a life expectancy of about three years at a working pressure of 5 bars. With increasing pressure the life expectancy is considerably reduced. However, the investigation showed that the life expectancy decreased less than linear with increasing pressure.

Long-duration tests on mechanical seals of water circulating pumps (6) have proved that life times of approx. 100,000 operating hours can be expected under certain constant operating conditions. Such extremely long service life will only rarely been reached in practice, because any increase of slide pressure, friction coefficient, slide velocity or temperature will increase

C603/033/2003 © IMechE 2003

the wear and thus shorten the life time. The temperature of the gliding faces has, apart from the influence of pressure, a decisive influence on the wear of the seal faces. A temperature increase can result in a reduction of stability, hardness or a disintegration of the impregnation of the carbon materials (7).

Early investigations of the influence of temperature on the wear of carbon materials (8) have revealed wear rates in the order of several hundred μm/h which increased drastically with increasing temperature. Today, the values of the wear usually found for mechanical seals are smaller by a factor of 100 to 1000 due to improvements of the carbon materials. However, the tendency of increasing wear with increasing liquid temperature can also be found in today's seals. Empirical tests on the life time of mechanical seals at different temperatures (9) have shown that life expectancies of 5 years at operating temperatures of less than 50°C can also be achieved in practice.

It becomes apparent from the previous discussion that both pressure and temperature have a large influence on the wear and thus the life expectancy of a mechanical seal. As mentioned at the beginning, other influences also play a role in the determination of the wear. Due to the complex mutual interactions of these influences and the pressure as well as the temperature it is very difficult to make a reliable forecast of the life expectancy based on theoretical predictions. It is thus necessary to determine the wear on a seal in experiments. For the use of a seal in practice the seal type, the used material combinations and the fluid are mostly fixed. Provided that the sliding velocity varies only slightly it will in such experiments be sufficient to vary pressure and temperature and measure the resulting wear. The disadvantage of this procedure is that the results cannot be easily applied to other types of seals, materials and fluids.

3   DIAGNOSTIC SYSTEM FOR MECHANICAL SEALS

3.1   Experimental details

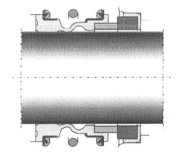

Figure 2. Sketch of the Rubber Bellow
Seal Used to Determine the Wear Rates.

All experimental data were obtained using a rubber bellow seal with a shaft diameter of 32 mm mounted into a commercially available pump for heating applications. A schematic drawing of the seal used is shown in Figure 2. The seal face was made of Antimony impregnated Carbon and the stationary seat was made of Silicon Carbide.

Each seal was tested at each operating condition for 100 hours. In order to determine wear rates at varying temperatures and pressures the height of the carbon was measured before and after the test run with a height measuring instrument with an accuracy of ± 0.5 µm. On the circumference of the seal face 4 points about 90° apart were selected and the height was determined by calculating the average value. After each test run the seal faces were re-polished. The mean roughness was checked to be below 0.1 µm and the deviation from an ideal flat surface was not bigger than 1 µm peak-to-peak. All experiments were carried out using deionized water with an electrical conductivity of 1 µS/cm.

3.2 Life time algorithm

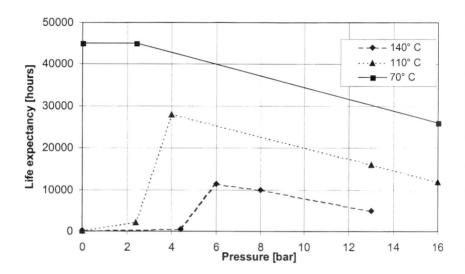

Figure 3. Experimentally Determined Life Time of a Mechanical Seal Depending on Pressure and Fluid Temperature.

The basis to determine the remaining life expectancy of a mechanical seal is the so-called life time algorithm which contains the detailed knowledge of relevant operating conditions as well as their influence on the wear of the seal. This influence has been experimentally determined at temperatures of 70°C, 110°C and 140°C and different pressures. The life times to be expected at such operating conditions are depicted in Figure 3. Note, that the life time shown in Figure 3 is the time after which the carbon face is worn down to a point that any further wear would lead to a significant leakage of the seal. Seal failure due to any other failure mechanism is not taken into account in the data shown in Figure 3. Further experiments to include other failure mechanisms are currently in progress.

As evident from the diagram the life expectancy is decreasing with rising pressure and temperature. This corresponds to the theoretical expectation depicted in paragraph 2. The steep decrease in life time at 110 °C and 140 °C at small pressures is due to the operation at a point below the vapor curve (high temperature and low pressure), leading to a very high rate

C603/033/2003 © IMechE 2003

of wear and thus a low life expectancy of the seal. From the knowledge of the expected life time at the three temperatures and different pressures the performance at other operating conditions can be deducted by linear interpolation. The functional relation of the life time as a function of temperature and pressure is integrated in the software of the diagnostic system and is the core of the life expectancy calculation.

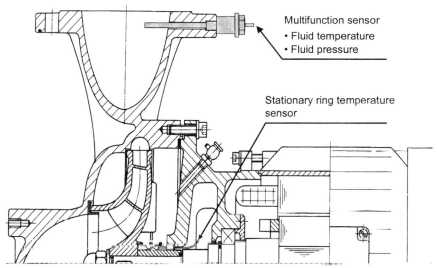

Multifunction sensor
• Fluid temperature
• Fluid pressure

Stationary ring temperature sensor

Figure 4. Cross Sectional View of a Glanded Pump with Integrated Sensors for the Seal Diagnostic System.

The fluid temperature, the pressure as well as the temperature of the stationary ring of the mechanical seal are continually measured during the operation of the diagnostic system. The temperature at the stationary ring is being utilized to detect dry-running at the seal by taking a temperature difference between the pumped fluid and the stationary ring of more than 50 K as an indicator for dry running of the mechanical seal.

A possible arrangement of the respective sensors is shown in Figure 4. The sensors for pressure and temperature of the fluid, integrated in a single housing, are mounted in the pump discharge flange. The temperature sensor is arranged in a way that the temperature is measured directly in the fluid stream. The temperature at the stationary ring is measured at a distance of approximately one millimeter from the sliding face of the mechanical seal thus ensuring a correct reading of the temperature of the sliding surface.

A wear factor is being attributed to each operating condition with the aid of the life time algorithm. The measured running time is weighted by the wear factor for each operating condition and then subtracted from the initial life expectancy. With this procedure it is possible to determine the remaining life expectance depending on the operating conditions run through in the past. Critical operating conditions, such as dry running, operating close to the boiling point as well as operation at excessively high fluid temperatures are treated in the same manner as normal operation. Each critical operation condition is attributed an accordingly high wear factor.

3.3 Warning at intolerable operation conditions

Whilst all the time periods in each operating condition are being summed to determine the remaining life expectancy any possible fault conditions (EFD) at the mechanical seal must be immediately notified to the management system (fault signals). The plant operator is thus enabled to initiate remedial measures and to rectify the actual fault causes for the seal failure. Maintenance and fault consequential costs are thus reduced and the safety of the plant increased.

3.3.1 Example: operations near the boiling point
In the operation of plants there are various reasons causing a pressure decrease or a strong increase of product temperature. For example defect expansion vessels in heating installations or insufficient inlet pressure can cause such a critical operating condition. This can lead to a partial evaporation of the product, resulting in insufficient hydro-dynamic lubrication of the seal faces and consequently in premature aging of the faces ('failure due to wear'). The boiling region as defined in the diagnostic system is shown in Figure 5. The distance between the two curves accounts for the fact that the fluid pressure is measured at the flange and not directly at the mechanical seal (see figure 4).

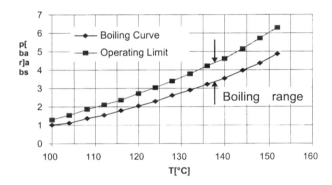

Figure 5. Implemented Alarm Limits of the Diagnostic System for Operations Near the Boiling Point.

3.3.2 Example: dry running of the mechanical seal
Mechanical seals must not be operated in dry-running conditions. The temperature sensor integrated in the stationary seal ring serves to detect such an intolerable condition. The characteristic temperature rise together with the fluid temperature as a reference value are used to indicate the dry running condition. In the experiments as shown in figure 6 dry running of the mechanical seal was deliberately induced repeatedly and in each case a drastic increase in temperature was detected at the integrated temperature sensor.

C603/033/2003 © IMechE 2003

For measurement of the seal temperature the temperature sensor is glued directly into the seal ring made from SiC. Thus, a reliable and accurate measurement of the seal face temperature can be carried through.

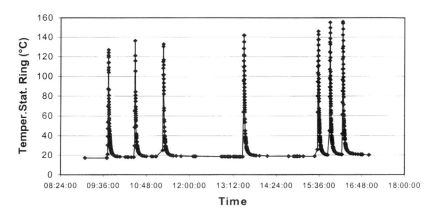

Figure 6. Temperature at the Stationary Seal Ring for Determination of Dry Running of the Mechanical Seal.

3.4 Scheme of Data Processing and Data Transmission

The table in Figure 7 shows a condensed version of how the values received by the sensor are registered, processed and displayed by the diagnostic system. The relays can be used to switch off the pump or to activate any remote signal according to the current status of the diagnostic system.

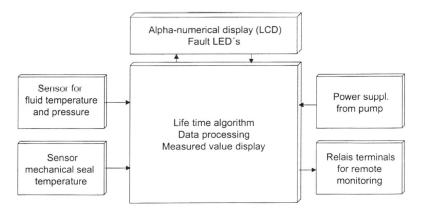

Figure 7. Scheme of the Diagnostic System with Sensor Inputs, LCD, LED's and Relay Outputs.

4   APPLICATIONS

4.1   Building services

Since March 2000 five diagnostic systems are installed in heating systems of big buildings, e.g. hospitals. The seal being used is a rubber bellows seal as shown in figure 2 for a shaft diameter of 32 mm. The maximum temperature is 140 °C and the maximum pressure is 16 bars. The liquid sealed in the field test is water with anti-corrosion additives. Deliberately induced and accidental fault conditions like operation in the boiling range were reliably detected. This led already in the field test to an optimization of the operation of a heating system avoiding such critical operation conditions.

4.2   Swimming pool pumps

Since 2001 about 200 units were brought in the field of swimming pool pumps. Here, the system is used as a dry running protection. The pump automatically switched off, when dry running is detected at the mechanical seal, thus avoid larger damage to location of installation of the pump.

4.3   Tool for warranty claims

Often both pumps and mechanical seals are used beyond the specified operating limits. The resulting failures are often difficult to investigate. The result are claims of the pump user towards the pump manufacturer and/or the seal manufacturer. In order to improve the basis of proof for all involved parties since March 2001 the system is used on ships as such a tool for warranty claims. The system is operated in a closed box and cannot be opened without breaking the lead seal. In case of a warranty claim the system is send in to the pump manufacturer without breaking the lead seal. By using the recorded history of operating data a judgement of the warranty claim is significantly facilitated.

4.4   Combined pump and seal monitoring

Since a mechanical seal represents only one component of a pump a combined monitoring system for the pump including the seal would be desirable. In a first approach to reach this target the diagnostic system for mechanical seals was combined with a system measuring vibrations at the pump installed in a refinery. The output signals of both systems are conducted to a central control room and combined into one common software. Therefore the operator gets a quick and easy overview over the condition of the pumps and seals at one glance.

5 CONCLUSION

Figure 8 shows a view of the diagnostic system for mechanical seals. All signals are led into the control and operating unit through PG cable glands. With the aid of the display the parameters necessary for the first run can be set and all memorized operating conditions can be readout menu-guided for a possible fault analysis. Red and yellow LED's indicate current fault conditions or an ending life time of the mechanical seal. The possibility of preventive maintenance (PM) as well as the continuous monitoring (EFD) of the mechanical seal will extend periods between maintenance and may partly avoid maintenance service calls, resulting in reduced life cycle costs and in an increased safety of the plant.

Due to the long life time of the mechanical seal under normal operation of several 10000 hours an end of the life time has not been reached yet. Therefore a final statement about the correct calculation of the remaining life time can not be made yet. Additional field tests in chemical plants and refineries are in preparation.

Figure 8. Glanded Pump with Integrated Diagnostic System for Mechanical Seals.

REFERENCES

(1)   Greitzke S. -F. and Strelow, G., 2000, "Reliability Considerations Applied to Serial Pumps- MTBF Analysis", Pump Users International Forum, Karlsruhe, Germany.

(2)   Flitney, R. K., 1987, "A study of factors affecting mechanical seal performance", Proc. Instn. Mech. Engrs. Vol 201, 17.

(3)   Archard, J. F., 1995, "Contact and Rubbing of Flat Surfaces", J. Appl. Phys. *24*.

(4)   Holm, R., 1958, *Electric Contact Handbook*, Berlin, Germany, third Edition, Springer Verlag.

(5)   Williams, J. G., 1965, "Shaft-Seal Systems for Large Power-Reactor Pumps", Nucleonics Vol. *23*, 2.

(6)   Mayer, E., 1968, "Gleitringdichtungen für Verbrennungsmotoren, elektrische Maschinen und Sondergetriebe, Konstruktion 20, 2.

(7)   Nau, B. S., 1990, "Research in Mechanical Seals", Proc. Instn. Mech. Engrs. Vol *204*, 349.

(8)   Johnson, R. L., Swikert, M. A., Bailey, J. M., 1956, "Wear of Typical Carbon-Base Sliding Seal Materials at Temperatures to 700°F", NACA Report 3595.

(9)   Flitney, R. K. and Nau, B. S., 1976, "Seal survey: Part 1- rotary mechanical face seals", BHRA Report CR1386.

C603/033/2003 © IMechE 2003

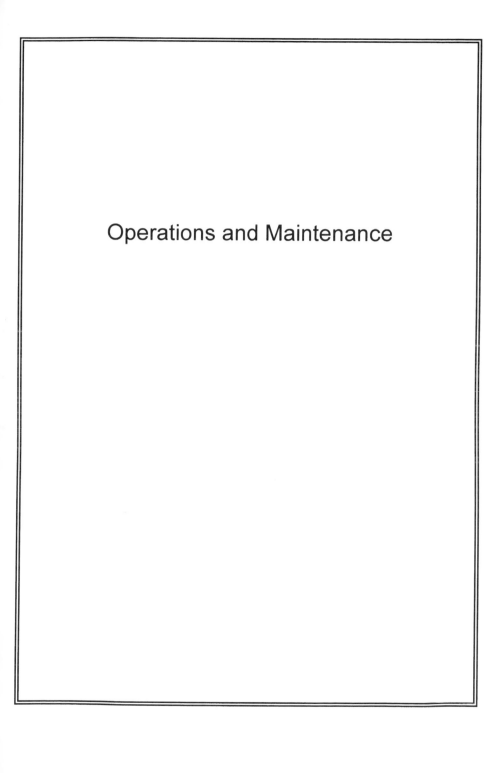

Operations and Maintenance

C603/004/2003

Unstable operation of a synthesis gas compressor

J P M SMEULERS and **IR O VAN WOLFSWINKEL**
TNO Institute of Applied Physics, Delft, Netherlands

1. INTRODUCTION

The compressor system of an ammonia synthesis plant showed unstable operation at part load conditions. Measurements showed that the instability developed at conditions far from the theoretical surge point. In spite of this the instability could drive the compressor into surge. By setting the anti surge control with a large surge margin the instability could be suppressed. As this implies that the anti surge by pass is partially opened and a flow is fed back from discharge to suction, this means a considerable reduction of the efficiency at part load conditions. In order to find other alternatives, a numerical model of the compressor system was made. With this model the effects of various optional modifications have been investigated. The model showed that the instability was caused by an interaction between the compressor and the pipe system. In this paper the model and some of the results are presented.

2. SYSTEM DESCRIPTION

The compressor system that has been investigated is part of a synthesis gas system of a fertiliser plant. The simplified layout is shown schematically in figure 1.

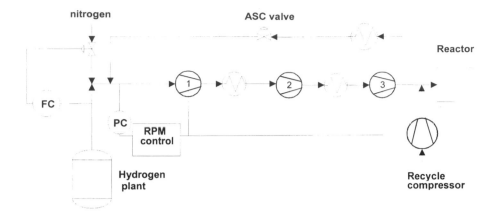

Figure 1. Schematic of the syngas compressor system.

The hydrogen plant is at a remote location and therefore suction line is relatively long. The interstage piping together with the volume of the heatexchangers represent a large volume. This affects the dynamics of the compressor.

The suction pressure is 23 bar and the discharge 133 bar. The three stages of the compressor each consist of a number of wheels so that each stage gives a pressure ratio of approximately 1.8. The compressor is protected against surge by means of an anti surge control system that opens a by-pass through which gas is recycled when the work point gets too close to the surge line.

The nitrogen flow is controlled such that the ratio between hydrogen and nitrogen is constant 3 to 1. The capacity of the compressor is controlled by means of a speed control on the suction pressure.

3. PROBLEM DESCRIPTION

The compressor was found to operate stable for the rated capacity. However, when the capacity of the plant was reduced the compressor was surging at a capacity far away from the surge point deduced from factory tests. This limited seriously the operating envelope of the compressor and therefore of the plant. A temporary solution was to apply a large surge margin so that the anti surge control keeps up the flow through the compressor. Though this proved to be an effective solution, the efficiency at part load conditions was reduced considerably. And therefore a better solution was searched.

4. MEASUREMENTS

In order to find the root cause of the early occurrence of surge measurements have been carried out at full and reduced loads. Dynamic pressure gauges have been installed at every available location in order to detect at which stage surge occurs the first time. However, the measurements showed an unexpected behaviour.

During the measurements the compressor was first operated on full capacity. Then the anti surge control was disabled (except for a safeguarding that acts when surge occurs) and the flow through the compressor was slowly reduced. During this procedure the pressures were recorded with a high sampling rate (typically 500 Hz) to be able to resolve high frequency events. The other process parameters such as flows and temperatures have been obtained from the DCS at a sampling rate of 4 Hz.

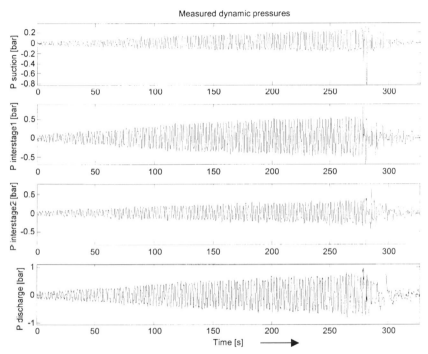

Figure 2. Part of a record of the dynamic pressures measured at the three stages of the compressor. Not all records are shown as at each stage suction and discharge side is measured. The capacity is slowly decreased and as a result the amplitude of the pulsations increases, which finally leads to surge.

A part of an interesting recording is shown in figure 2. From this it can be seen that low frequency pulsations occur even at full capacity. When the capacity is reduced there is a point where the amplitude of the pulsations is not constant anymore, but start to grow autonomously. Finally the compressor gets into surge, indicated by the sharp pressure peaks. The by pass is immediately opened by the second level anti surge controller to bring the compressor out of surge.

In figure 3 is zoomed in on a part of the records shown in figure 2. The pressure variations appear to be almost sinusoidal and have a frequency of ca. 0.4 Hz.

From this it was concluded that the surge phenomenon is not a classical surge phenomenon. The surge is a result of large pressure variations (pulsations) in the system. Apparently these pulsations are generated by a system instability. The fact that the capacity of the compressor is an important condition for the occurrence of surge is an indication that the compressor is part of the instability.

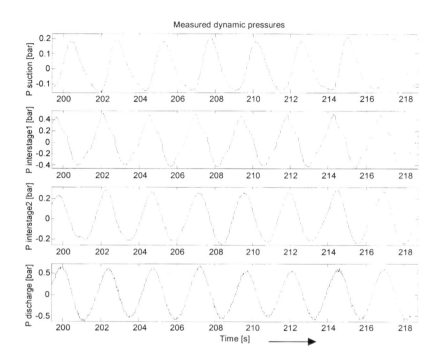

Figure 3. Zoom in on a part of the records shown in figure 2.

A comparison of the pressure differences across (or the head of) each stage reveals an interesting phenomenon. It appears (see figure 4) that each stage shows a phase shift of approximately 60 degrees, giving a total phase shift of 180 degrees over the compressor. Clearly the compressor and the interstage piping act as an active element in an oscillator. The way in which the oscillator works has been analysed by means of a simulation model of the compressor and the pipe system. This is described in the next section.

C603/004/2003 © With Author 2003

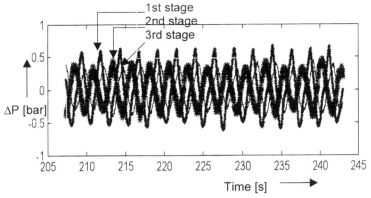

Figure 4. Comparison of the pressure differences across the three stages of the compressor.

5. THE SIMULATION MODEL

A simulation model has been built using the PULSIM program[1]. The program is a building box with various types of elements such as pipe sections, volumes, 1-branches, (control) valves and also models of fluid machinery. Each element models the flow and pressure in the element as a function of time. There is also a simulation element that represents a stage of a turbo compressor. A model of the compressor is composed by coupling of three compressor simulation elements with elements that represent the flow in the connected pipe sections, vessels and heatexchangers.

A special element is the controller with which the low level process control can be modelled. The program calculates the pressures and flows in the system as a function of time for all points of interest. At will various disturbances can be applied to the system.

The model used to investigate the problem presented above includes the hydrogen plant, e.g. the volume of the last vessel that delivers the hydrogen to the compressor, the nitrogen line starting at the control valve, the suction line up to the compressor, the compressor, and the discharge line up to the reactor. The compressor was modelled in detail. Each of the three stages was modelled by the head-flow characteristic of that stage. Also the interstage piping and the heatexchangers have been modelled in order to represent the correct phase shift.

6. RESULTS AND MODIFICATION OF THE SYSTEM

In the simulation model the system appeared to be stable for the entire capacity range. However, a small disturbance of the suction pressure caused a relatively large variation of the flow, showing that the system was marginally stable. In order to find the governing

---

[1] The PULSIM program has been developed by TNO Institute of Applied Physics for the modelling of pressure pulsations in pipe systems and fluid machinery.

frequencies a study was made of the response to a sinusoidal flow disturbance in the suction with a varying frequency. In figure 5 the amplitude of the response is shown as a function of the frequency for the suction first stage, the discharge first stage, the discharge second stage, and the discharge third stage. It appears that peak in the response occurs at a frequency of 0.5 Hz, approximately the measured frequency of the instability.

The response of the first stage is the strongest and is decaying in the higher stages. In the measurements the reduction of the amplitude was far less. The simulation results show a phase shift between the stages that is similar to what has been measured.

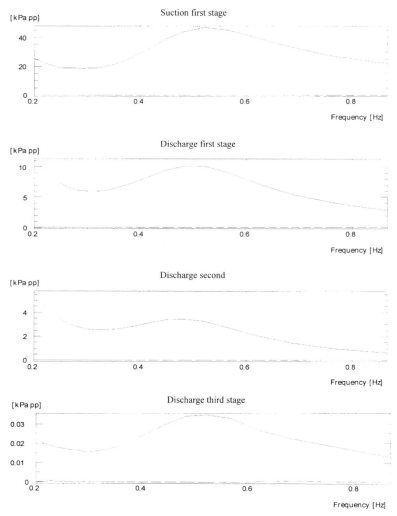

Figure 5. Response of the simulation model to a sinusoidal excitation with a frequency varying between 0.2 and 0.9 Hz.

An explanation for this can be that there are additional effects, which are:
1. The rotational inertia of the rotor and the variation of the rotation speed. Assuming that the driving power stays constant, e.g. the speed control is much slower than the phenomenon, the speed will vary due to variations in pressure difference and mass flow. Due to the inertia of the rotor the rotation speed lags the variation of the load. When the pressure difference increases the speed will (after some time) decrease and the compressor characteristic will change accordingly. This effect has not been accounted in the calculated response. As all stages are on one axis, variations in one stage are more or less copied to the other stages, causing similar pressure variations like in the other stages. During the measurements a variation of the speed has been observed. Therefore it can be concluded that variation of the rotation speed is also affecting the stability and the dynamic behaviour.
2. Another effect that can undermine the stability of the system is the variation of the molecular weight of the gas that enters the compressor. Caused by a flow variation the ratio of hydrogen and nitrogen flow may vary as a result of pressure and flow variations. It is assumed that the nitrogen flow is constant as the flow controller has a longer response time than the instability cycle. The time constant of this phenomenon is the travel time of a change of composition from the mixing point to the compressor, which is typically the transport time of the flow. As this travel time is of the order of half the period of the instability, this effect may contribute to the instability.
3. The control of the suction pressure can be unstable. It is assumed in general that controller actions are much slower than the 0.4 Hz response. Therefore it is not likely that this effect will contribute to the unstable behaviour.

The effects discussed above will delay the response of the system and therefore the frequency of the instability will reduce to 0.4 Hz. The resonance peak at 0.5 Hz in the calculated response curves of the pipe system is still considered to be the main cause of the instability. Therefore changing this response is therefore the main task to control the problem. In order to remove the instability the measures described below have been considered.

Eliminate or dampen the resonance.
The simplest way is to dampen the resonance by means of an orifice plate in the suction line. In order to minimise pressure loss (and loss of efficiency) instead of an orifice plate, a control valve can be used that is opened during full load and is partially closed when the compressor is running at partial load. Simulation runs show that this is an effective way to control the problem.

Installation of a large vessel in the suction line is another possibility. Such a vessel dampens the flow variations and in this way also the fluctuations of the suction pressure. Besides also the variations of the molecular weight of the gas are reduced.

Steeper compressor characteristics
By changing the head-flow characteristic of the compressor the impedance of the compressor for pressure variations can be changed. By making the characteristic steeper a pressure variation will cause a smaller flow variation. However, changing the compressor characteristic is very difficult to achieve, especially in an existing installation.

Improve control
The phenomena considered are fast in terms of the time constants of the process control, however, not that fast that it cannot be controlled by means of conventional control loops. There

are three points where improved control can help. First the control of the suction pressure can be changed to faster control of the suction flow. The setpoint of the flow controller can be regulated by means of the suction pressure. This way of cascaded control will especially be more stable for the low capacity range of the compressor where a small change of the pressure difference across the compressor will lead to a large change of the flow.

The control of the nitrogen supply can be made faster so that molecular weight variations are suppressed. It is expected that this will have a small effect as the molecular weight variations have a limited effect on the stability of the compressor and should be considered as an additional measure.

A last possibility is to improve the speed control of the compressor. When the speed controller on the driver counteracts the speed variations that are caused by the variation of the flow and the pressures. It will be a rather difficult task to implement a speed control that is fast enough to achieve this as the driver and the compressor have a large inertia.

7. CONCLUDING REMARKS

The case history described in this paper shows that the application of a centrifugal compressor in a process installation needs the careful investigation of the dynamic behaviour of the compressor in interaction with the pipe system. In this case the long suction line and the large volume of the interstage and discharge system create a good condition for a resonance.

For this installation various options could and have been investigated by means of a dynamic flow model of the compressor and pipe system. To describe the complete dynamic behaviour of the system also the rotor inertia of the compressor and the driver should be included as well as the controllers. In view of the time available these effects have not been investigated extensively in this case.

8. REFERENCES

1. Pampreen, R.C., "Compressor surge and stall", Concepts ETI Inc., ISBN 0-933283-05-9
2. Cumpsty, N.A., "Compressor aerodynamics", Longman Scientific&Technical UK, ISBN 0-582-01364-X
3. Fink D.A., Cumpsty N.A., Greitzer E.M., "Surge dynamics in a free-spool centrifugal compressor system", Journal of Turbomachinery, Vol. 114/321-332, April 1992.
4. Sparks, C.R., "On the transient interaction of centrifugal compressors and their piping systems", ASME 83-GT-236
5. Willems, F.P.T, "Modelling and bounded feedback stabilisation of centrifugal compressor surge", PhD thesis University of Technology Eindhoven, June 2000.
6. Botros, K.K., "Transient phenomena in compressor stations during surge", Journal of Engineering for Gas Turbines and Power, 116, 133-142
7. Smeulers, J.P.M., Bouman, W.J., van Essen, H.A., "Model predictive control of compressor installations", Proceedings of the Conference on Compressors and their systems, I Mech E, London 1999.

C603/012/2003

Maintenance strategy development for new liquefied natural gas (LNG) facilities

N ARTHUR
KBR Production Services

ABSTRACT

This paper describes the development of optimised maintenance activities for new fluid machinery and ancillary equipment installed as part of the expansion of existing L iquefied Natural Gas (LNG) facilities. This maintenance development ensures that the Health, Safety, Environmental (HSE), production and cost performance of the asset will be assured throughout the lifecycle of the plant, hence providing the lowest possible cost of plant ownership.

1.   INTRODUCTION

A Kellogg Brown & Root (KBR) consortium was awarded a US$1.5 billion, lump sum contract to execute a major expansion of an LNG complex. The consortium includes a number of companies and is known collectively as the Joint Venture (JV). When complete, the LNG complex will be the largest LNG facility in the world and the expansion increases the p lant p rocessing c apacity from 1 5.8 t o 2 3 m illion metric tons of LNG / annum. This requires the installation, operation and maintenance of high numbers of capital-intensive fluid machinery.

The scope of work for the facilities includes design, procurement, construction, and commissioning of two LNG trains and offsite facilities to be added to the existing six-train facility. Feedstock for the trains, which liquefy natural gas for storage, is the natural gas from reserves some 100km offshore. The LNG will be shipped via super tankers to existing and new LNG buyers, where it will be re-gasified and piped to customers.

The first train of the new facilities is scheduled for completion during the fourth quarter of 2002, and the second train during the third quarter of 2003. In each phase, this LNG complex has incorporated state-of-the-art technologies, featuring air-cooling throughout the facilities, employment of very large gas turbines, and a high-efficiency design that maximizes the utilization of the turbine's power.

KBR Production Services (KBR-PS) were contracted to carry out a Maintenance Strategy Development (MSD) for these new facilities. The MSD was completed as it is imperative that the life-cycle operation and maintenance costs of this equipment are minimised in order to maximise plant productivity and asset performance.

The scope of work included all equipment items for the new facilities except that of a new slug catcher, metering and instrumentation. The slug catcher and metering were excluded from the scope of work as the operator assumed responsibility, and instrumentation maintenance was covered by through an Instrument Protective Function (IPF) study.

Traditionally, the MSD for new build assets occurs later in the project phase, and as such, the potential impact on the lifecycle costs of the asset is rarely considered in detail. In order to minimise operational lifecycle costs, it is fundamental that any MSD work is considered early in the project phase. There are significant benefits to carrying out the MSD work at this stage of the project, namely;

- Lowest possible lifecycle cost of asset ownership.
- Detailed manning and planned outage information available for plant start-up.
- Implemented Maintenance Management System (MMS) for plant start-up.
- Established continuous improvement processes for immediate implementation.

This paper describes how the MSD for the expansion of the new LNG facilities was carried out, and gives an overview of the corporate processes and procedures involved. The results of the MSD are provided along with the further work necessary to transform these results into a fully functioning MMS.

2   MANAGEMENT SYSTEMS

The KBR Production Services Management System (PSMS) defines a number of corporate business processes. One of these business processes (1) refers to the way that the management of maintenance and inspection activities is carried out. This business process encompasses the definition of maintenance strategy and planning, implementation, performance management and the management of change.

The process for maintenance management defines and manages those maintenance activities required to safeguard technical integrity and meet the HSE, production and cost performance business needs of the asset in the most cost effective manner, over the life cycle of the asset. The process for maintenance management is applicable to all KBR-PS operations and is implemented at all locations, taking due cognisance of any contractual and regional requirements.

The rigorous application of this process ensures that a consistent approach to all maintenance activities and provides a clearly defined means of conducting, measuring, auditing and delivering maintenance work. Whilst these processes are well defined, they remain flexible enough to accommodate innovation and continuous improvement. This approach allows the effective and efficient generation of maintenance for KBR-PS customers as is described in the following sections of the paper.

3.   MAINTENANCE STRATEGY DEVELOPMENT

Each element of the MSD for the new LNG facilities is briefly covered in the following sections of the paper. An overview Process Flow Diagram (PFD) of the facilities is shown in Figure 1 and details the main 'units' or systems of the plant.

3.1   Criticality analysis
The first stage of the MSD was a criticality analysis of the equipment items constituting the new LNG facilities. The criticality analysis was carried out to;

- Assess the impact of the failure of each of the units and sub-units of the new LNG facility in terms of HSE, production and repair consequences.
- Rank the equipment of the facility in terms of their financial criticality.
- Identify the criticality of the equipment terms of prescribed criticality bands.
- Identify the most appropriate MSD technique based on criticality.

Each of the units constituting the new facilities was decomposed into a number of 'sub-units' for the purposes of the criticality analysis. This was done as smaller groups of equipment items could be dealt with, and critical equipment could be identified more specifically. In doing this the 32 units were sub-divided into 64 sub-units. A criticality analysis was then carried out on these 64 sub-units. The criticality of each of the LNG sub-units was estimated by calculating the product of the *probability of failure* of each sub-unit with the *consequence of failure* of that sub unit.

3.1.1   Probability of failure
The probability of failure of each sub-unit was estimated by considering;

- The total amount of equipment in the sub-unit and available redundancy.
- The size, power and complexity of the equipment in the sub-unit.
- KBR-PS experience of similar equipment, operational situations and relevant failure rates.

3.1.2   Consequence of failure
The consequence of failure was estimated in financial terms using experience of typical comparable equipment operating in oil & gas operations using the following four criteria;

- HSE consequences.
- Deferred LNG production.
- Equipment repair costs.

This allowed each of the new LNG sub-units to be ranked in terms of annualised financial criticality. This criticality is the potential risk, or commercial exposure, that can be ascribed to the failure of each of the plant sub-units. Clearly, those sub-units of relatively high criticality require the most intensive effort in terms of MSD. This ensures that the customer gets best value from the effort expended.

3.1.3 Criticality bands

All major equipment within facilities have been specified according to three generic definitions (3), those definitions are (in descending order of criticality); vital services, essential services and non-essential services. In addition to determining the criticality of each of the sub-units, each sub-unit was banded within one of the criticality bands defined above.

3.2 MSD techniques

When determining optimised maintenance for equipment, three methods of MSD are commonly used. Each of these MSD techniques has concomitant advantages as will be described in the following sections.

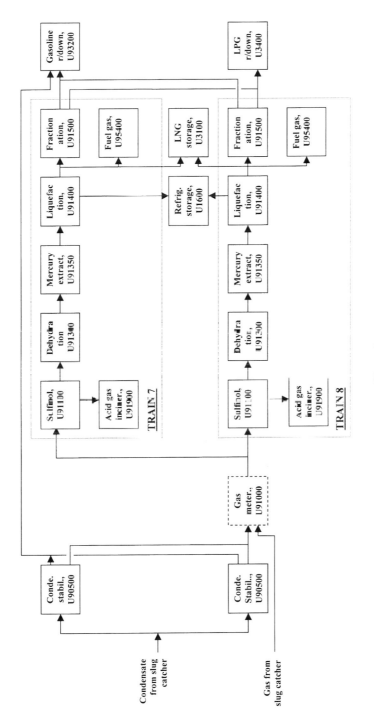

Figure 1

3.2.1 Detailed Reliability Centred Maintenance

Detailed Reliability Centred Maintenance (RCM) is a process used to identify specific optimised maintenance activities for equipment items. This maintenance is detailed and tailored explicitly to the individual equipment item. The detailed RCM was carried out according to KBR-PS best practice methodology (1) and industry-recognised standards (2). The main elements of the detailed RCM work were;

- Failure Modes and Effects Analysis (FMEA).
- Failure Characteristics Analysis (FCA).
- Maintenance Strategy Selection (MSS).

Typically, detailed RCM is only applied to equipment items of high criticality. This ensures that high MSD effort is applied only where necessary and where maximum benefits will be accrued.

3.2.2 Read across RCM

Read across RCM was applied to equipment identical to those treated by detailed RCM. The detailed RCM analysis and maintenance recommendations were exploited and applied to these identical equipment items. In this way the technical rigour of the detailed RCM approach was retained and applied to these identical equipment but at a fraction of the initial effort. Read across RCM is applied to equipment items of criticality similar to that of detailed RCM.

3.2.3 Generic Reliability Centred Maintenance

Generic RCM involves the determination of generic equipment categories and conduction of RCM at a generic equipment level, rather than at a specific equipment level. For example, some equipment categories that were covered through the generic RCM approach were Low Voltage (LV) motors, hand valves and gear pumps of low power. KBR-PS owns and maintains a massive suite of generic RCM libraries. These libraries harness the learning of the organisation and represent best-in-class generic maintenance.

3.3 Maintenance activities

In general, maintenance activities applied to equipment items on the new LNG facilities can be divided into four main sub-categories;

- Overhaul maintenance activities
- Condition Based Maintenance activities.
- Failure finding maintenance activities.
- On-failure maintenance activities.

Each of these categories will now be discussed in turn.

3.3.1 Overhaul maintenance activities

Overhaul maintenance activities are performed on equipment where failures are age related and are 'clock resetting' activities based on a finite, predictable wear out characteristic of the equipment. The periodicity of overhaul maintenance activities recommended as part of the MSD is at the maximum interval, whilst ensuring that all HSE and production consequences of failure are mitigated.

3.3.2 Condition Based Maintenance activities

Most equipment failures are not age-related and are therefore not mitigated effectively by the application of overhaul maintenance activities. Many failures develop gradually (2) over a period of time with the rate of failure being random. In these instances, Condition Monitoring (CM) activities can be used to detect the onset of failure and to instigate remedial maintenance to avoid that failure a nd to m itigate t he consequence o f failure i .e. C ondition Based Maintenance (CBM). The use of CBM has been maximised in the MSD.

3.3.3 Failure finding maintenance activities

A hidden failure is a failure which on it's own will not become evident under normal circumstances, and hence will have no direct consequences. A simple example of this is a seized bearing on an auxiliary lubrication pump. The seized bearing does not become evident until the main lubrication pump fails, and the auxiliary unit is required. Hidden failures are mitigated through failure finding maintenance activities. Failure finding maintenance activities do not alter the inherent failure rate of the equipment, however, they do identify that a failure has occurred within that equipment. The frequency of failure finding maintenance activities is governed by the desired availability of the equipment. By determining a target availability of the system and by having access to accurate failure data allows an appropriate testing interval to be determined.

3.3.4 On-failure maintenance activities

With every maintenance activity described in the preceding sections comes an associated cost. That cost comprises various elements such as labour, specialist equipment and spare parts. For certain equipment items the actual cost of carrying out maintenance of whatever type is in excess of the benefits gained from carrying out the maintenance i.e. the commercial benefit of carrying out the maintenance activity is less than the a ctivity c ost. I n these instances, on-failure maintenance was applied and a maintenance activity takes place only when equipment fails. The repair activity is that required to return the failed equipment item to service. On failure maintenance activities are generally applied to low value equipment whose failure has no impact on HSE or production.

4. RESULTS AND DISCUSSION

4.1 Criticality analysis

Of the 64 sub-units analysed in the criticality analysis, 6 were identified as being of 'vital' criticality, 5 of 'essential' criticality and the remaining 53 of 'non-essential' criticality. This is typical of such studies, with a high criticality concentration in a small number of equipment items. The most critical sub-unit identified was that of Fire Protection. This is unsurprising as this sub-unit has a significant HSE consequence of failure. Table 1 shows the g eneral distribution of MSD technique with sub-unit criticality that was applied in the MSD.

Table 1

Strategy type	Equipment category		
	Vital	Essential	Non-essential
Detailed RCM	X		
Read across RCM	X	X	
Generic	X	X	X

Reliability Centred Maintenance Worksheet

Project :	LNG
System :	Power Generation
Main tag	U-94050

	FMEA					MSS			FCA
Function	Function Failure	Failure Reference	Tag Number(s)	Failure Mode	Local Effect	System Effect	Maintenance Strategy Ref	Maint Task Reference	Maintenance Tasks	Interval	Failure Characteristic Analysis and interval selection criteria
To safely and efficiently generate 11kV 20MW 60Hz electrical power.	Does not generate electrical power.	940.01.A.0 1	G-94050	Generator rotor deteriorates (all generators).	Deterioration of the rotor (all failure results in generator failure. Alarm initiated to alert the operator.	No and rotor power to user. end electrical	C-OCM	E.01.01. 01.E2	With the generator running, carry out analysis and trending using in-built and/or portable Condition monitoring systems. Compare with previous analysis and where results indicate unacceptable deterioration, schedule repairs or re-schedule condition monitoring.	104	REF: H B&R Template E.01.01 FCA data. Category E2
		940.01.A.0 2	G-94050	Generator winding high / low resistance (all generators).	Generator running currents become low (all unbalanced, unit user. protection operates and shuts down the generator. Alarm initiated to alert the operator.	No power to end electrical	C-OCM	E.01.01. 02.E2	Carry out the measurement, analysis and trending of the generator winding resistance. Compare with previous results and where results indicate unacceptable deterioration, schedule repairs or re-schedule condition monitoring.	104	
		940.01.A.0 3	G-94050	Generator winding low insulation resistance (all generators).	Generator windings damaged by fault current. Unit protection operates and shuts down the generator. Alarm initiated to alert the operator.	No power to end electrical	C-OCM	E.01.01. 03.E2	Carry out the measurement, analysis and trending of the generator winding insulation resistance. Compare with previous results and where results indicate unacceptable deterioration, carry out a polarisation index and if polarisation index is below desired schedule repairs or re-schedule condition monitoring.	104	REF: H B&R Template E.01.01 FCA data. Category E2
		940.01.A.0 4	G-94050	Generator condensation heater anti- insulation	Unit anti- protection operates and shuts down power the generator user. heater, no generator	No power to end electrical	C-OCM	E.01.01. 04.E2	Carry out the measurement, analysis and trending of the generator anti-	104	REF: H B&R Template E.01.01 FCA data. Category E2

Item	Equipment	Failure mode / cause	Failure effect	Type	Task No.	Task description	Interval	Reference
940.01.A.0 5	G-94050	Generator open circuit (stand-by generators). resistance (stand-by generators). condensation heater - heating when required. Generator damaged by condensation.	Generator fails to run when required. Operations staff would be unaware of this failure during normal operations. No electrical power to end user.	E=FFT	E.01.01.05.E2	Start and run the generator, check that the generator runs satisfactorily. condensation heater insulation resistance. Compare with previous results and where results indicate unacceptable deterioration, schedule repairs or re-schedule condition monitoring.	104	REF: H B&R Template E.01.01 FCA data. Category E2
940.01.A.0 6	G-94050	Generator bearing wear due to degradation of lubrication (stand-by generators).	Noise, vibration and temperature increases. Bearing seizes, unit operates. Alarm and shuts down the generator. Alarm initiated to alert the operator. No electrical power to end user.	C-OCM	E.01.01.06.E2	Take lube oil sample and analyse compare and trend analysis results schedule and carryout lube oil replacement as requirement is identified.	13	REF: H B&R Template E.01.01 FCA data. Category E2
940.01.A.0 7	G-94050	Generator bearing failure due to other failures (binding, seizing, scoring). (duty generators)	Noise, vibration and temperature increases. Bearing seizes, unit operates. protection operates and shuts down the generator. Alarm initiated to alert the operator. No electrical power to end user.	C-OCM	E.01.01.07.E2	Carry out a vibration survey, analysis and trending of the generator. Compare with previous results and where results indicate unacceptable deterioration, schedule repairs or re-schedule condition monitoring.	5	REF: H B&R Template E.01.01 FCA data. Category E2
940.01.A.0 8	G-94050	Generator bearing failure due to excessive wear, brinelling/fretting (duty generators).	Bearing collapses, unit protection operates and shuts down the generator. No electrical power to end user.	C-OCM	E.01.01.08.E2	Carry out vibration analysis and trend results where analysis results indicate the requirement, schedule the removal of old bearings, inspect housings and shaft, record findings, re-instate with new bearings. Carry out base vibration readings.	5	REF: H B&R Template E.01.01 FCA data. Category E2
940.01.A.0 9	G-94050	Generator bearing damage due to standby run for limited period (stand-by generators).	When started, the generator will run with increasing bearing noise, vibration and temperature. Bearing unit collapses, protection operates and shuts down the generator. Alarm to alert the operator. No electrical power to end user.	C-OCM	E.01.01.09.E2	Stand-by machines carry out pre start checks run machine ensure correct running conditions are achieved, check for excessive noise, temperature or vibration, on completion return to to stand-by condition.	13	REF: H B&R Template E.01.01 FCA data. Category E2

Criticality: Partial Results

Unit	Unit description	Sub-unit	Sub-unit description	Criticality (US $)	Classification
96000	Fire Protection	96000.3	Fire Prot. - Gas Detection	6,237,489	Vital
96000	Fire Protection	96000.4	Fire Prot. - Active Fire Prot.	6,216,060	Vital
96000	Fire Protection	96000.2	Fire Prot. - Fire Detection	6,216,055	Vital
91400	Liquefaction	91400.2	Liquef. - Cool & e/flash	5,531,975	Vital
91400	Liquefaction	91400.3	Liquef. - MR comp	5,531,975	Vital
91400[2]	Liquefaction	91400.4	Liquef. - Propane	5,531,975	Vital
96300[1]	Pressure relief & liquid disposal	96300.1	Pres. Rel. & liq. disp. - Wet flare	2,388,849	Essential
96300[1]	Pressure relief & liquid disposal	96300.2	Pres. rel. & liq. disp. - Dry flare	2,388,849	Essential
96200	Telecommunication	96200.2	Telecom. - Radio System	2,244,229	Essential
91500	Fractionation	91500.1	Fractionation - Demethan.	2,129,456	Essential
91500	Fractionation	91500.2	Fractionation - Deprop. & debutan.	2,129,456	Essential
94000[5]	Electricity Generation	94000.1	Gen. - Primary	1,179,575	Non-essential
91100[6]	Sulfinol	91100.3	Sulfinol - Regeneration	1,074,374	Non-essential
91100	Sulfinol	91100.1	Sulfinol - Feed & absorb	1,064,728	Non-essential
91400	Liquefaction	91400.1	Liquef. - Scrubbing	1,064,728	Non-essential
3100	LNG Storage & Loading	3100.2	LNG Stor. & Load - Stor	1,017,301	Non-essential
96000	Fire Protection	96000.1	Fire Prot. - Firewater	742,176	Non-essential
3100	LNG Storage & Loading	3100.4	LNG Stor. & Load - Load	528,267	Non-essential
91900	Acid Gas Incinerator	91900.1	A/gas incin. - A/gas KO	405,378	Non-essential
91900	Acid Gas Incinerator	91900.2	A/gas incin. - F/gas KO	405,378	Non-essential
91900	Acid Gas Incinerator	91900.4	A/gas incin. - Oxa atomise	405,378	Non-essential
94000	Electricity Generation	94000.3	Gen. - com. trans.	405,378	Non-essential
95000	Electricity Generation	95000.1	Gen - Train trans.	405,378	Non-essential
91100	Sulfinol	91100.2	Sulfinol - Circulation	401,366	Non-essential
90510	Condensate Stabilisation	90510.2	Cond. Stab. - Recomp.	224,427	Non-essential

4.2   MSD application

4.2.1   Detailed Reliability Centred Maintenance
Detailed RCM was carried out on equipment in 4 of the 64 sub-units owing to the high criticality of these sub-units. The sub-units that were considered using detailed RCM were;

- Sub-unit 94000.1 Power generation – Primary. See attached worksheet.
- Sub-unit 91400.2 Liquefaction – Cooling and end-flash compression.
- Sub-unit 91400.3 Liquefaction – Mixed refrigerant compression.
- Sub-unit 91400.4 Liquefaction – Propane compression.

4.2.2   Read across Reliability Centred Maintenance
Read across RCM was applied at an equipment and sub-unit level in order to maximise efficiencies.

4.2.3   Generic Reliability Centred Maintenance
Generic RCM was used to treat all equipment items not considered by detailed or read across RCM. This meant that minimum development of generic maintenance was required and that the maintenance c ould be developed f or t he o verwhelming m ajority o f t he n ew equipment very quickly.

4.3   Maintenance activities
For each equipment item an optimised combination of the four types of maintenance activity were prescribed as discussed previously. The combination of maintenance activities were optimised to ensure the HSE and production business requirements of the asset could be delivered at minimum lifecycle cost of ownership

4.4   Continuous Improvement
For every individual equipment item that was part of the MSD, KBR-PS have provided JV with the following information;

- Full FMEA including fully referenced data sources.
- Recommended maintenance tasks, durations and periodicities.
- Man power and trade requirements.
- Spares and special tools requirements
- Any assumptions made.

The provision of this level of detail makes an ideal platform to ensure continuous improvement in the maintenance effectiveness and efficiency of the f acilities. A s w ith a ll green field development, the initial maintenance implemented is based on assumption and historical knowledge of similar facilities. In order to ensure that the maintenance is continually aligned with the HSE, production and cost business targets of the asset, continuous i mprovement p rocesses and tools must be implemented. In this way changing commercial, plant configuration and reliability data can be factored and the asset maintenance model re-run for optimum performance.

5.   CONCLUSIONS AND FURTHER WORK

In conclusion it may be seen that KBR-PS have carried the efficient and effective MSD for the new LNG facilities. This has been done by exploiting corporate best practices and harnessing the lateral learning of previous, similar projects. A number of different MSD techniques have been discussed and their application dependant on criticality.

The MSD has ensured that the facilities has available an optimised maintenance strategy designed to meet health, safety, environmental and business needs at the lowest possible lifecycle cost. The preceding sections of this document have described the procedures and methodologies employed in the derivation of optimised maintenance tasks and frequencies for the equipment.

HKBR-PS have delivered detailed, recommended maintenance to JV, however, it is recognised by both parties that further work is required to transfer this into a fully functioning MMS prior to plant start-up. This is the subject of ongoing discussion between HKBR-PS and JV and further reinforces the benefits if carrying out this work at the early stage of projects.

REFERENCES

(1)   KBR-PS Business Process PM-GL-BR-MAI-002, 'Maintenance and Inspection Strategy Development'.
(2)   J. Moubray, 'Reliability-centred Maintenance', 6th edition, Butterworth-Heinemann Ltd., 1991.
(3)   KBR-PS document 'Memorandum of understanding between JV and HKBR-PS covering RCM based maintenance evaluation', 26th January 2001.

ABBREVIATIONS

KBR	Kellogg Brown and Root
KBR-PS	KBR Production Services
FCA	Failure Characteristics Analysis
FMEA	Failure Modes and Effects Analysis
HSE	Health, Safety and Environmental
IPF	Instrument Protective Function
JV	Joint Venture
LNG	Liquefied Natural Gas
LV	Low Voltage
MMS	Maintenance Management System
MSD	Maintenance Strategy Development
MSS	Maintenance Strategy Selection
O&M	Operations and Maintenance
PFD	Process Flow Diagram
PSMS	Production Services Management System
RCM	Reliability Centred Maintenance

© KBR Production Services.

C603/018/2003

Operational difficulties on an ethylene plant cracked gas compressor train

A D R CHAPMAN
BP Chemicals, Sunbury-on-Thames, UK
G M CORRIGAN and **B R GAMBLIN**
BP Chemicals, Grangemouth, UK

SYNOPSIS

This paper will describe the operational difficulties encountered on the auxiliary systems serving a cracked gas compressor on an ethylene plant at the BP complex at Grangemouth, Scotland. The compressor is a three casing, five-stage machine of 25 MW capacity driven by an extraction/admission/condensing steam turbine. The compressor and turbine, which were both manufactured in Europe, were first commissioned in 1993; they are shown in figure 1.
The problems covered will include excessive seal oil make up, excessive sour oil leakage, seal oil/lube oil cross contamination, buffer gas control, the effects of oil contaminated with process gas and problems with the wash water system.

The paper will also detail the actions taken both on line and off line to understand the root causes and then to implement the appropriate remedial work.

1 SYSTEM DESIGN

1.1 Ethylene plant compressors
Due to the nature of the process ethylene plants are in continuous service between planned maintenance outages with run times that are measured in years, typically four to six. The compressors used in these plants are usually single train and, because they are vital for production, the operators' expectation is that they will have one hundred per cent availability between the maintenance outages.

An end user will buy an ethylene plant design from an engineering contracting company that has the technology rights. Although some plant designs have compressors on other duties all ethylene plants have an ethylene refrigeration compressor, a propylene refrigeration compressor and a cracked gas compressor (sometimes referred to as a charge gas compressor). This last duty takes the mixture of gases from the cracking furnaces and compresses it to a sufficiently high pressure to allow the gas separation processes to proceed.

This paper is concerned with a cracked gas compressor on an ethylene cracker within the Olefins Group of plants at the BP Complex, Grangemouth, UK. The cracker was first commissioned in 1993.

1.2 Turbo-compressor system design

With such a demanding availability required of the machinery the detailed specifications of the turbo-compressor systems are based around recognized national, or international standards, such as those of the American Petroleum Institute (API), to give the end user a measure of assurance.

1.2.1 Compressor

The compressor is of European manufacture and its design was based on the fifth edition of API617: Centrifugal Compressors for Petroleum, Chemical and Gas Service Industries. It has three casings in a single shaft line: a low pressure (LP), a medium pressure (MP) and a high pressure (HP). There are five compression sections in all: a double suction, single discharge first section contained in the LP casing; sections 2 and 3 are arranged discharge to discharge in the MP casing and sections 4 and 5 are similarly arranged in the HP casing.

The original duty was saturated cracked gas at 35 kPa(g) at the first section suction, fifth section discharge pressure of 3400 kPa(g), first section mass flow of 120 tonnes/hour, molecular weight of 21 kg/kg-mole, and a rated power of 23 MW. In 1996 the compressor was re-rated for a flow of 134 tonnes/h at a power of 25.4 MW.

1.2.2 Turbine

Manufactured in Europe, the turbine is an extraction / condensing type built to API612: Special Purpose Steam Turbines for Petroleum, Chemical and Gas Industry Services, with a facility for admitting steam via the extraction steam header during the start up of the plant.

The steam conditions are 10 000 kPa(g), 500°C at inlet, 4200 kPa(g) at the extraction steam header and condensing at 20 kPa(a) giving a turbine power of 25.4 MW.

Turbine control is by a redundant, electronic governor with speed as the prime controlled variable. Extraction steam pressure is also a controlled variable but the turbine is operated to maximize the 10 000 kPa(g) steam usage therefore, either the inlet steam header low pressure limit or the high extraction steam flow limit is always active before the extraction pressure set point is achieved.

1.2.3 Couplings

The compressor train has three couplings in accordance with API671 Special Purpose Couplings for Petroleum, Chemical and Gas Industry Services. In addition each one has a torque measuring capability (based on measuring the angle of twist of the coupling spacer) from which the power transmitted from shaft to shaft is calculated.

1.2.4 Lube oil and control oil system

The lube oil and turbine control oil system is designed to API614: Lubrication, Shaft Sealing and Control Oil Systems and Auxiliaries for Petroleum, Chemical and Gas Industry Services and comprises the usual features of a stainless steel oil reservoir (12 m^3 capacity), a turbine driven, positive displacement main oil pump, a motor driven, positive displacement standby oil pump, duplex coolers, duplex filters, accumulators and direct acting pressure control valves. Piping is all stainless steel.

The pressure levels are 1000 kPa(g) for control oil and 150 kPa(g) for lube oil.

1.2.5   Seal oil system
While the seal oil system is separate from the lube and control oil system, it has the same components and design standard.
The pressure levels controlled are the seal oil header at 1700 kPa(g) and three differential pressures (one for each compressor casing), each maintaining 250 kPa across the mechanical contact seals within the compressor casings.
A sour seal oil recovery system comprising six degassing traps means that the gas is led back to the compressor first section suction drum and the liquid is drained into a heated degassing tank before spilling back to the main seal oil reservoir.

1.2.6   Buffer gas system
Buffer gas is a non-fouling gas injected into a labyrinth seal positioned inboard of the main mechanical contact seals on each compressor casing. While the buffer gas was initially a stream recycled from the fifth section discharge, it was later changed to a mixture of ethylene and methane to ensure cleanliness.

1.2.7   Lube oil and seal oil separation within the compressor casings
The design of the compressor shafting and bearing housings is such that there is a labyrinth seal between the lube oil return and the sweet seal oil return. This labyrinth is purged with nitrogen at a nominal 25 hPa(g) pressure.

1.2.8   Anti-surge system
Since a key protection for any centrifugal compressor is an anti-surge system, the five compressor sections of this compressor are protected by a total of three recycle lines. A dedicated, proprietary electronic controller is used on each of the three anti-surge loops.

2   OPERATIONAL HISTORY

This section describes some of the operational difficulties encountered with the cracked gas compressor since its first commissioning.

2.1   Unbalanced first section flows
2.1.1   The symptoms
The LP compressor casing handles the first section of compression and is a double suction, single discharge design, which has the advantage of a balanced thrust load. There is also an anti-surge recycle valve protecting this casing.

When the machine was first commissioned on recycle it was noticed that there was a discrepancy between the gas temperatures measured in the suction drum, one of the suction pipes below the compressor and the discharge piping. The temperature in the suction pipe was in excess of 90°C and was similar to the discharge pipe temperature. The suction drum gas temperature was the expected value of 35°C.

An investigation revealed that the compressor was only taking suction gas through one of the two inlets and was not only discharging the gas via the discharge pipe but also down the second inlet pipe. It was fortunate that the only suction temperature transmitter was installed in the pipe taking the reverse flow otherwise the recycling flow might not have been noticed until more damage was done.

2.1.2 The remedy
Having identified what was happening the cause was not immediately apparent; especially since the process indicators within the plant distributed control system (DCS) and emergency shutdown system (ESD) were showing valves to be in the correct position.

While theories such as "rotating stall" were suggested the answer lay in a physical check of the positions of all of the valves in the process flow paths. When the recycle valve opening was found to be actually much less than was indicated by its telemetry, the fault was corrected. As a result the compressor flows and temperatures changed to the expected directions and values and the next stage of commissioning was started.

2.2 Unstable oil systems
2.2.1 The symptoms
Although the main oil pumps on both the lube/control oil system and the seal oil system are driven by steam turbines, the stand-by pumps are motor driven and are set to start automatically given certain events e.g. a low oil pressure.

In order to check the operability of the stand-by pump a routine is performed to switch over pumps. During the factory acceptance testing of the oil system this check was done satisfactorily as part of the API614 requirements, although in this case the turbine driver was replaced with a slave motor because the factory did not have a steam system.

At site the pump switch-over from the turbine driven to the motor driven was done successfully but the reverse switch caused the turbine governor to become unstable. This led to the turbine tripping out on overspeed and the motor driven pump cutting back in.

2.2.2 The remedy
Since the control valves were of a self-acting design (to give a fast speed of response) and the turbine governor was a hydraulic type there was limited scope for adjusting the tuning of the systems.

A dynamic simulation was commissioned, using a university with close links to the original equipment manufacturer (OEM), and video footage of the turbine tachometer was also used to demonstrate the unstable oscillations of speed up to the trip point and to validate the model results.

The simulation suggested several solutions ranging from the installation of more oil accumulators to alteration of the turbine dynamics. The preferred option was the installation of a turbine rotor with a much higher mass moment of inertia that returned the system to a stable state.

C603/018/2003 © IMechE 2003

2.3 Vanishing seal oil (cracked gas mousse)
2.3.1 *The symptoms*
The plant operators were reporting that they needed to add fresh oil into the seal oil reservoir so that a low level alarm could be cleared. Although a long-term trend of the oil level did show a saw tooth pattern as the oil was replenished, there was no sign of oil leakage around the system.

Elsewhere in the process other odd things were happening in drums and separator vessels.

2.3.2 *The remedy*
The seal oil system is designed to recover the sour oil passing the mechanical seals by first separating the gas from the oil in traps then passing the oil into a holding tank which is heated to drive off the remaining gas before returning the oil to the main seal oil reservoir.

The flow of gas and sour oil from the compressor casings is determined by the size of a restriction orifice in the vent line from the trap. It turned out that the sizes of the orifices fitted to the six traps (one per compressor seal) were too large and the venting gas was entraining sour oil and taking it back to the process drums where it formed a mousse. The orifice sizes were reduced and the mousse production was resolved.

2.4 Lube and seal oil transfer at the compressor casings
2.4.1 *The symptoms*
Having apparently solved the 'vanishing seal oil' problem, it was a surprise when the operators reported that they were still adding oil. A review of the long-term trends of level in both the seal oil reservoir and the lube/control oil reservoir showed that the seal oil reservoir level was falling while the level in the larger lube/control oil tank was rising. Since the operators were not topping up the lube/control oil system, there must be some transfer of oil occurring.

While the two oil systems were in close proximity within the bearing housings of the compressor casings, the lube oil chamber was separated from the sweet seal oil chamber (i.e. seal oil that has not come into contact with process gas) by a labyrinth seal which also had a nitrogen purge applied.

It was believed that this was the site of the oil transfer and it was also noticed that the oil transfer could go from lube oil to the seal oil and vice versa.

2.4.2 *The remedy*
On the basis that there was a pressure differential between the two chambers in the bearing housing two pressure balancing holes were drilled at the top of the labyrinth carrier walls at each end of the compressor casings. Unfortunately this did not stop the transfer.

To date the transfer mechanism is not understood and so no remedy is in place. Instead the transfer is tolerated and managed by inter-reservoir oil transfer as dictated by the tank levels.

2.5 Excessive steam usage at re-starts following trips
2.5.1 *The symptoms*
It is well known that cracked gas is a fouling duty and the performance of the compressor will deteriorate the longer it is in service after an overhaul.

After a couple of years service the operators noticed that the barring gear failed to turn the shaft line. (With steam turbines of a certain bearing span a barring device is often installed to allow for an even cooling of the shaft given a trip. This prevents thermal bowing of the rotor and consequent high vibrations on start-up.)

With the barring gear out of action the actions following a trip were to restart the machine and hold at 500 rpm for a period of time until the vibrations were settled. As time went on the amount of steam taken to restart the compressor set increased; at one time 26 tonnes/h was recorded.

2.5.2   The remedy

It was suspected that fouling was responsible for the failure of the barring gear to turn the shaft line. At the next overhaul of the machine this was proved to be the case with severe fouling discovered behind the impeller covers that acted like brake pads on a car.

Since it was also established that there had been an oxygen excursion during the operation and this had promoted the excessive fouling found actions were taken to prevent a reoccurrence.

Up to that point there had been no wash water employed in the compressor although an injection header had been provided by the OEM. It was decided to utilize this water injection system following the overhaul and process water was to be the selected medium.

In addition it was decided to apply a proprietary anti-fouling coating to the compressor rotors and to the diaphragms. This was engineered by the OEM and coating manufacturer and was installed in 1996.

2.6   Excessive sour oil leakage from the seals
2.6.1   The symptoms

The design leakage rate for the mechanical contact seals used on the shafts is 200 ml/h per seal and these rates were checked during the acceptance testing at the factory. Post test inspection showed some remedial work was needed to the seal oil distribution ring and to some of the clearances so that heat removal could be improved.

The seals, manufactured by a European company, are mechanical contact type with a rotating face of silicon carbide running against a carbon face held in a bellows.

For a period of months after the first commissioning the sour oil leakage rates were within the design rates, which were measured by blocking the sour oil drain from the trap and recording the time taken to fill up the sight glass (a known volume).

When the sour oil rates began to increase, particularly from the larger diameter seals in the LP and MP casings, the run down lines to the sour oil traps were warm to the touch and samples of the sour oil from the MP casing sometimes contained water.

The rate of sour oil leakage continued to increase until the recovery system could not properly de-gas the oil and sour oil was returned to the main sweet seal oil reservoir. At its peak the combined sour oil leakage rate from the six compressor seals was five hundred litres an hour – more than four hundred times the design flow from six seals.

2.6.2   The remedy
No remedy could be identified until the seals could be inspected and replaced with the spare parts during the next planned overhaul. Until then the leakage rate had to be tolerated.

A close watch was kept on the oil transfer between the lube/control system and the seal oil system and as long as the transfer was from the lube oil to the seal oil then the operation was accepted.

To remove the water from the sour oil rundowns the source of the buffer gas was changed to provide a warmer and drier gas to the seals.

Other checks such as monitoring of the flash point of the seal oil and also the viscosity of the oil were invoked to ensure continued safe operation. The viscosity of oil contaminated with hydrocarbon will reduce but, as the duty was sealing and not lubrication, this was considered less critical. A lower limit for viscosity was defined as acceptable for the seal oil duty and a programme to continually change out the contaminated seal oil for fresh oil was established to maintain the measured viscosity above this limit.

2.7   Blocked filters on the electro-hydraulic converters
2.7.1   The symptoms
The machine train suffered a number of trips due to blocking of the inlet strainers on the electro-hydraulic converters used in the turbine control system. These strainers are in-line, 10 micron screens at the inlet to the converters. Analysis proved that there was sulphur present in the material that was blocking the strainers and investigations proved that bubbling process cracked gas through the oil resulted in sulphur being precipitated out into the oil. This precipitated sulphur was actually sub-micron in size and therefore should not have blocked the strainers. Further investigations found another substance that appeared to be binding the sulphur particles together and this proved to be the anti-foam agent from the oil. This material was found to have formed in all the filters within both the seal and the lube and control oil systems.

Problems were also encountered at each of the globe valves in the sweet seal oil return lines; the same material was forming within the valve and partially blocking the flow. It would then clear and cause a dip in the differential pressure between seal oil supply and reference gas at the seal resulting in an alarm or, in one or two cases, a machine trip.

2.7.2   The remedy
The filtration of the control oil supplied to the turbine was improved by the installation of two large filters in series, a 6 micron element followed by a 3 micron, to replace each of the existing single 10 micron filters.

The electro-hydraulic converter screens were removed and replaced by an external cartridge filter set-up comprising differential pressure monitoring and a spare filter.

The grade of oil used in the system was changed to remove the anti-foam additive to prevent further build-up of this additive in the bulk oil during the on-going change-out of seal oil to maintain viscosity. The anti-foam additive had been precipitating out of the oil and the continual addition of fresh oil had increased its concentration further exacerbating this problem.

A regular programme to open and reset the sweet seal oil return line control valves was established to keep them clear of material.

Subsequently a thorough programme of cleaning and flushing of the whole oil system was planned for the next major overhaul of this machine train.

2.8 Wash water injection

2.8.1 The symptoms

Wash water was installed on this machine in 1996 to alleviate the problems experienced with polymer fouling. This wash water system was run from 1996 to 2000 without any operational problems. The first four stages were supplied with process water from the compressor first stage suction drum and the fifth stage was supplied from the fifth stage discharge cooler.

When the compressor was opened up in 2000 for overhaul it was discovered that significant erosion and corrosion had occurred within the first four compressor stages. The inter-stage seal clearance areas were significantly enlarged, corrosion cracking was found within some rotor impellers and under impeller shaft corrosion had occurred. Most significant was that the shaft end labyrinth seals had been corroded right through and were loose on the shaft (Figure 2). Buffer gas could not be supplied at a high enough pressure and so the process cracked gas could not be prevented from contacting and contaminating the seal oil. The loss of these labyrinths and the subsequent failure to segregate the process gas was found to be a primary cause of the problems described in sections 2.6 and 2.7. The fifth stage was in pristine condition.

2.8.2 The remedy

After a thorough inspection and analysis of individual components the compressor was rebuilt and endorsed for a limited operating period by the OEM (Figure 3).

Investigations attributed the problems to the acidity of the wash water supplied to these four stages. Wash water was therefore commissioned to the fifth stage only until a chemical dosing system had been installed to maintain a suitable pH of the wash water supply to the first four stages.

A material upgrade for the rotors and diaphragms was agreed with the OEM with the new components to be installed at the end of the reduced operating period.

3 CONCLUSIONS

3.1 Review the system design

When problems occur the fault analysis can concentrate on what has changed since the last steady state. The experience of the authors has been that an additional review of the system design can be essential to finding the root cause of the problem. In other words check how the system was designed to operate, challenge the design and, if it is correct, ensure operation to that design.

3.2 Understand the systems

Many systems are now supplied as packages and during factory acceptance testing and on site commissioning adjustments are made to the systems to iron out problems. As things wear or off-ideal operational conditions are encountered it is vital that any changes to the systems are made by those who understand the systems thoroughly. These changes must be implemented via a strictly applied "control of change" procedure that includes the key steps of documentation and training. Such a procedure must span the numerous phases of a project from initial design, through construction and commissioning and then into operation.

3.3 Monitor system operation to be within design

Prudent monitoring of system operation is key if faults or problems are to be picked up early and their impact minimized. This includes short and long term trending, using all instrumentation (both locally installed as well as that feeding the control room screens of the plant distributed control system), setting acceptance criteria for the parameters monitored from samples (e.g. oil viscosity, wash water pH). The actions to be taken following any alarm should also be clearly defined.

Figure 1 Cracked gas compressor train.

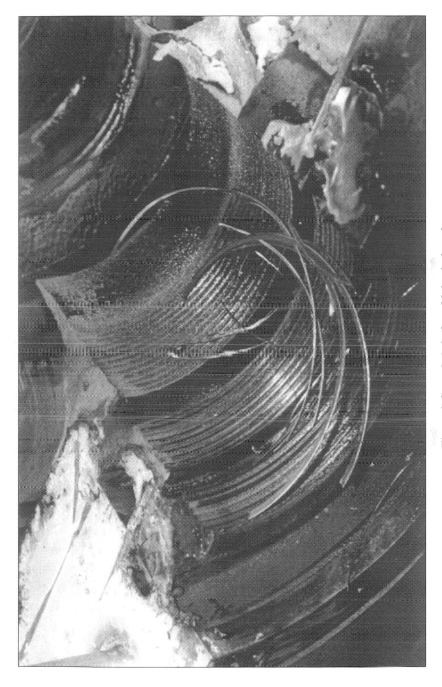

Figure 2 Corroded shaft end labyrinth seals.

Figure 3 Repaired MP rotor & diaphragms

C603/018/2003 © IMechE 2003

C603/034/2003

Water injection system to maintain the performance of cracked gas compressors in ethylene plants

T LOSCHA and **H MAGDALINSKI**
Siemens PGI, Duisberg, Germany
M GRAFE
Veba Oel Verarbeitung, Gelsenkirchen, Germany

Abstract:
Polymerisation inside the compressor leads to fouling, results in lower compressor efficiency and effects higher discharge temperatures, which again increase the polymerisation rate. As critical threshold for polymerisation a temperature of app. 90°C is normally considered. A lower discharge temperature can be reached either by reducing the pressure ratio, which is limited by the process itself, or by injecting water between the impellers.

The constantly injected water between the impellers evaporates. Due to the evaporation heat consumed the discharge temperatures decrease. The cooling effect also reduces the head necessary for the specified pressure ratio, so that the speed is reduced by 1 to 2% in general.

This paper summarises the physical background, construction requirements and experiences from the installation of a water injection system within cracked gas compressors with the aim to reduce the process gas discharge temperatures preventing polymerisation.

Prior to the selection of the water injection system, thermodynamic calculations to determine the discharge temperatures, injection quantities, pressures and power requirements are to be made. Also the material suitability of the compressor itself as well as of the heat exchangers, separators and auxiliaries has to be checked.

Based on these results the injection system including nozzles has been developed together with VEBA OELVERARBEITUNG as operator. The offered design for the water injection nozzle carriers is successfully used in the VEBA plants.

The thermodynamic calculations as well as the experiences of the operators show that the process gas discharge temperatures can be decreased by injecting water and fouling is reduced. This results in longer terms of operation and higher efficiency than without water injection.

1 INTRODUCTION

1.1 General

Measures against polymerisation can be divided into washing systems to remove polymer coatings and measures preventing polymerisation. Washing systems inject solvents occasionally to wash deposits away. As a high discharge temperature encourages the polymerisation reaction, measures preventing polymerisation are based on lowering the discharge temperature. One possibility is the decrease of the pressure ratio, i.e. the increase of the suction pressure and / or decrease of discharge pressure, which is limited by the process itself. Another possibility is the injection of water to be evaporated and thus giving a cooling effect on the process gas leading to a lower discharge temperature. As the cooling effect is significant, the power required for compression is normally reduced.

1.2 Benefits from the installation of a water injection system

Polymerisation inside the compressor leads to fouling, resulting in lower compressor efficiency and higher discharge temperatures. The main benefits from the installation of a water injection system within a raw gas compressor are the prevention / lowering of polymer formation deposits inside the compressor resulting in:
- higher efficiency
- lower discharge temperatures
- longer continuous operation periods

2 THEORETICAL BACKGROUND

2.1 Polymerisation

Polymerisation is favoured by the existence of long chain hydrocarbons and the existence of catalysts. Furthermore, temperature has a major influence on the reaction velocity. The polymerisation reaction velocity increases significantly above approximately 90°C.

Polymers have a consistency similar to tar, this is tough and sticky at high temperatures to be found in compressors.

In raw gas compressors, polymers are formed preferably at locations with high temperature at which they are deposited. Special attention has also to be paid to the labyrinth ends of the impeller inlet, the balance piston and the inserts. The labyrinth clearances become nearly completely overgrown. The surfaces of the rotating parts are normally coated up to approximate 2 mm, at higher thickness the coating flake off due to centrifugal forces.

Smooth surfaces can minimise the formation of polymers. Therefore, another preventative measure additionally to water injection, is the application of "anti-fouling-coating" on surfaces circulated by gas. The "anti-fouling-coating" effects very smooth surface avoiding / minimising deposit of polymers and thus fouling.

2.2 Water evaporation

The continuous injection of water between the impellers effects cooling of the process gas by evaporation. The evaporation heat of water is 2046 kJ/kg at 8 bar and thus significantly higher than just the heat capacity.

C603/034/2003 © IMechE 2003

To obtain an optimum heat transfer and hence an optimum cooling effect the surface of the water drops related to the volume has to be large, i.e. the water drops have to be small. The difference pressure between water injected and process gas should be in the range of 7 to 8 bar in our experiences. Correct injection guarantees that the water injected is evaporated completely.

The water to be injected has to meet at least the following requirements:
- boiler feed water quality
- low oxygen content as oxygen operates as catalyst for the polymerisation reaction

2.3   Effects on compressor (theory): head, power consumption

The water injection has the following main effects on the thermodynamics of compressor:
- Due to the cooling effect the head required for the compression is lowered.
- The mass flow through the compressor is increased effecting higher head required.
- The average molecular weight is decreased effecting higher head required.

In general, the cooling effect on lowering the compression head is higher than the head increase by mass flow increase and molecular weight decrease. That way, the head required is decreased in total. Thus, the speed and the power consumption are decreased.

The thermodynamic calculations for optimal water injection are normally based on the following assumptions:
- process gas temperature at outlet should be as low as possible, thus at the process stage outlet 100% of the water is evaporated
- injection pressure of water app. 7 8 bar above process gas pressure at least

The injection process itself can be divided in three steps: injection – atomisation - evaporation. In the injection nozzle, the pressure is converted into velocity. By injecting the water into the process gas stream, the water is atomised and immediately vaporised. To get an evaporation within short time, good heat transfer conditions are required for which the surface area of the drop related to the volume is important. A smaller surface effects a better heat transfer.

3   DESIGN

3.1   General

Fouling inside the compressor effects, at constant speed, the head-volume curve and the efficiency are decreased. At constant process pressure the suction volume flow is decreased. If the raw gas compressor is driven by a steam turbine, there is the possibility to increase the speed to increase the suction volume flow to the original value. But, the disadvantages are that the efficiency is lower and the discharge temperatures are higher leading again to increased polymerisation. The speed adjustment is limited by the maximum allowable speed and the maximum available steam. Thus, the operation period between two shut downs is decreased.

A typical raw gas compressor / cracked gas compressor consists normally of two or three casings, i.e. low pressure casing (LP), middle pressure casing (MD) and high pressure casing (HP).The following cross section drawing shows a typical raw gas compressor divided into LP- / MD- and HP-casing:

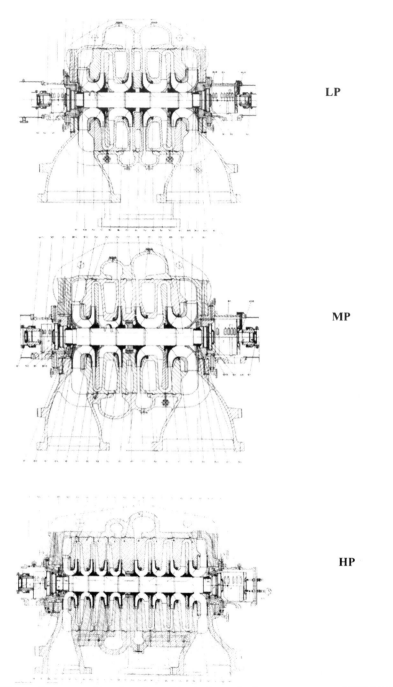

LP

MP

HP

Figure 1-cross section drawing of typical cracked gas compressor (LP/MP/HP casing)

C603/034/2003 © IMechE 2003

3.2 Construction requirements for raw gas compressor

The following is required to retrofit a compressor with a water injection system:

- nozzles for water injection system available at compressor casing
- nozzles for drainage available at compressor casing
- material of compressor suitable for wet conditions (H_2S)

3.3 Construction requirements for auxiliaries

The materials of the auxiliary system have to be suitable for wet conditions. Furthermore, the capacity of the intercoolers has to be checked regarding the amount of water to be separated. Additionally all capacities of down stream equipment, for example separators and steam traps, have to be examined

3.4 Requirements for water to be injected

The water to be injected should be of boiler feed quality, i.e. desalinated and de-mineralised. In addition to this, the oxygen content should be low (catalyst for polymerisation reaction). The water to be injected should be filtered to approximately 200 μm. The temperature should be in the range of 40°C to 50°C. To get a good atomisation, i.e. small drops the pressure difference between water and process gas has to be 7 to 8 bar at minimum.

3.5 Thermodynamic design

At first, the amount of water to be injected is determined by thermodynamic calculations. In general, the calculation basis are the process data and the condition, that the process gas is 100% saturated after each process stage. Second, the nozzles are chosen, the design of the nozzles devices and the water injection system including a tank, pumps, flow meters, control valves and piping is defined.

Main results of the thermodynamic calculations are:

- amount of water to be injected
- expected discharge temperatures
- power required
- speed
- performance envelope for new conditions

In the following the thermodynamic results for water injection into four process stages of a raw gas compressor are summarised.

The amount of water to be injected at each individual impeller is shown in figure 2. Each of the four process stages consists out of a certain number of impellers. The moisture content at the impeller inlet without water injection is shown over the inlet pressure (points). The curves describe the saturation for the individual process conditions at the impellers / stages. The process gas is saturated at the inlet to each process stage as the gas is cooled and water separated between the process stages. The amount of water to be injected is described by the moisture difference between the individual process operating point and the saturation curve (in detail shown for process stage I in the figure). Additionally, the amount of water to be separated by the intercoolers is shown as example for stage II. The amount of water to be separated in the intercoolers is with water injection higher than without water injection.

The calculated discharge temperatures without and with water injection are shown in figure 3. The average temperature decrease is 17°C approximately for this example, so that the discharge temperatures are below 91°C. With water injection the power at the coupling of the compressor and the speed are slightly lower than without water injection (app. 1%).

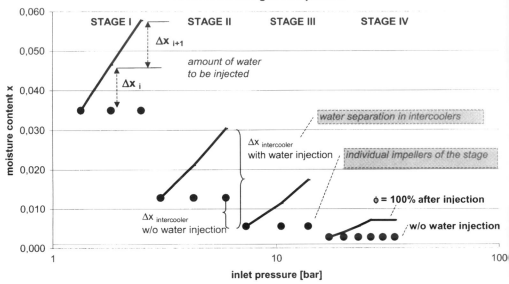

Figure 2 - amount of water to be injected

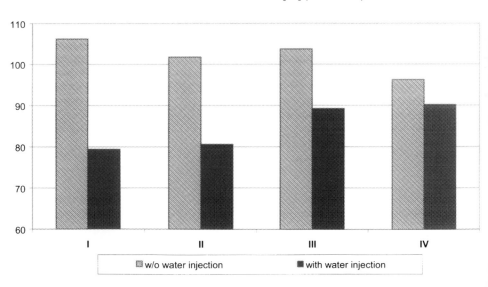

Figure 3 discharge temperatures calculated with and w/o water injection

C603/034/2003 © IMechE 2003

3.6 Nozzle carrier device

Our current design of the nozzle carrier device is based on experiences gained by Siemens PGI (formerly Demag) and VEBA as end-user of the water injection system, it is used in VEBA plants.

The design allows the change of injection nozzles during operation, if for any reason an injection nozzle is clogged, for example by suspended particles in the water or formation of polymers during interrupted water supply.

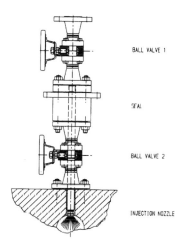

Figure 4 nozzle carrier device

The nozzle carrier device consists mainly out of the nozzle carrier, a ball valve, a sealing package and another ball valve screwed onto the compressor casing. Furthermore, a fitting/remove device is used for nozzle change.

For removing of a nozzle the following main steps are required:
closing of the ball valve (1), depressurising and drainage of the piping system, remove of piping connections, degassing, mounting of fitting/remove device, opening of ball valve (2) (screwed onto casing), remove of carrier device just that the ball valve (2) can be closed again, closing of ball valve (2), withdraw of nozzle carrier device. The fitting of the ball valve is in reverse order. In this way, the exchange of the nozzle during operation is ensured .

3.7 Design of the water injection system

The water injection system includes typically pumps, control valves, valves, flow meters, temperature and pressure measures, piping and if needed a tank. All piping, valves and equipment should be made from stainless steel.

During start up and shut down, it has to be ensured that the water supply is stopped to avoid any collection of water inside the compressor. Before start-up the drainage system has to be opened, to drain off any water collected.

In the figure below, a sketch of a typical water injection system is shown:

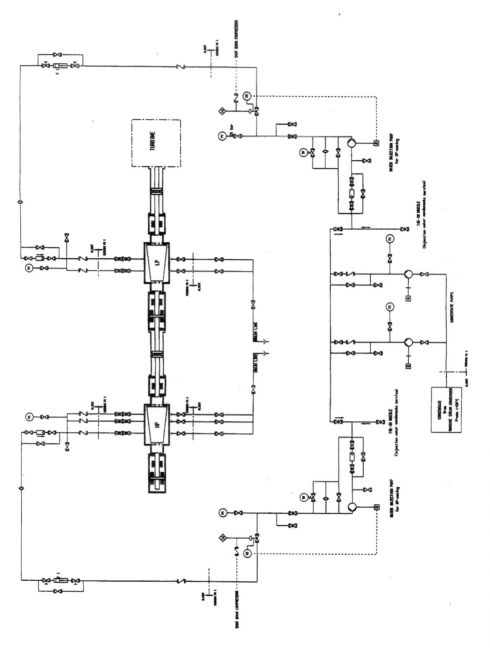

Figure 5 sketch of typical water injection system

4 OPERATION EXPERIENCES / REFERENCES

On Gelsenkirchen site VEBA operates two ethylene plants. Olefine 3 from 1972 and Olefine 4 from 1992. Each plant has one cracked gas compressor.

The Olefine 4 - compressor was equipped with a water injection system from the beginning. The injection fluid is boiler feed water for stage 5. Stage 1 to 4 are cooled with condensate from the separators located after the process heat exchangers in the cracked gas compressor system. During operation a slightly reduction of efficiency was noted. After nearly 6 years the casings were opened for the first time to inspect the internals and clean the machine as necessary.

The experience with the Olefine 3 - compressor without water injection, was that a cleaning stop was necessary every after 30 month of running. The discharge temperatures for the Olefin 3 plant are at the start and end of run are listed in the following table:

Olefin 3 (without water injection): gas temperatures						
Stage		1	2	3	4	5
Start of run (2 January 98)						
Suction temperature	[°C]	34	38	26	40	24
Discharge temperature	[°C]	86	94	92	100	61
End of run (6 July 02)						
Suction temperature	[°C]	31	40	35	35	27
Discharge temperature	[°C]	92	100	100	96	65

The following picture was made during revision of Olefin 3 during July 02 and shows the deposits in the MP raw gas compressor (without water injection):

Figure 6 Olefin 3 MD compressor without water injection – deposits at impeller

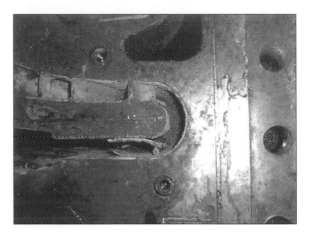

Figure 7 - Olefin 3 MD compressor without water injection – deposits at return channel

The Olefine 4 - machine was not clean, but the amount of polymerisation deposits was not comparable to the Olefine 3 - machine. The flow cross-sections were in comparatively good condition. De-mounting of the diaphragms and diffusers was difficult because of the sticky dirt in the fits. The cleaning could be done with glass-bead blasting, because the coat of polymers was not as thick as in the Olefine 3 - machine.

5   SUMMARY

The thermodynamic investigations show that the discharge temperatures can be decreased significantly by injecting water.

VEBA experience shows that the deposits in the cracked gas compressor with water injection system are significantly lower than in a comparable compressor without water injection, which indicates a significant reduction of process gas discharge temperatures.

The design of the injection system, including nozzle carriers, was developed together with Veba as operator and is successfully used in the Veba plants.

In general, it can be concluded that the installation of a water injection system effects a significant decrease of polymer deposits. This results in long term operation without maintenance at higher efficiency.

Retrofits and Upgrades

C603/011/2003

Major rerate of a process air compressor during a plant shutdown

G KERR
DuPont Polyester Technologies, Wilton, UK
F G CATALDO
DuPontSA, Wilton, UK

SYNOPSIS

The process air for a Polyester Intermediates plant was provided by the main air compressor (20MW) and a smaller subsidiary compressor (4.5MW). The small compressor was reaching the point where further operation would require significant expenditure without any performance improvements. This led to the plant seeking an alternate source of additional process air. Alongside this the process conditions had changed over time such that the main compressor was running significantly away from its original design point. Different options were assessed and a decision taken to uprate the main process air compressor. The operating constraints meant that the modification had to be designed such that construction and testing could be completed within the three week plant shutdown.

This paper describes the design and engineering applied through the project process to minimise the risk involved with such major changes. To ensure all aspects were considered the project team included personnel from the operator, the technical consultant, the detailed design contractor, the vendor, and the construction contractor.

The major challenges were around producing a suitable design capable of being installed and tested within the shutdown. The requirement was for an increase in mass flow rate of 20%, which combined with a reduction in discharge pressure equated to a 30% increase in discharge volume. This necessitated a new HP compressor casing which had to be mounted onto the existing bearing housings, internal modifications to the LP compressor and a larger first stage intercooler. The reuse of the existing bearing housings meant that it was not possible to carry out conventional API testing of the new components at the vendor's works before installation.

1   EXISTING CONFIGURATION

Two centrifugal air compressors provided the process air for a Pure Terephthalic Acid Plant. The main plant air compressor (T8 PAC) was commissioned in 1981 but subsequent plant improvements have meant that it is unable to supply all the process air requirements. In 1987 a compressor (T6 PAC) from a redundant plant was recommissioned to provide about 20% of the current air demand. This compressor was one of a pair originally commissioned in 1964 and uprated in 1968 by installing a suction booster fan. The total process air supply to the plant is 150,000m³/hr. The original delivery pressure was 31bara at a temperature of 190°C. At the time of this project the required delivery pressure had dropped to 24bara.

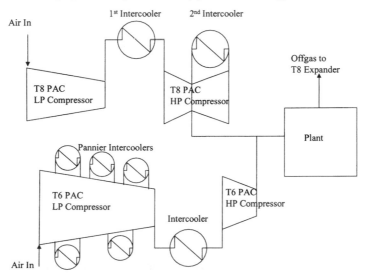

Figure 1 Existing process air supply to plant

The T6 PAC had been mothballed in 1981 and then recommissioned in 1986 but this work did not include any replacement of the original equipment. At a routine inspection in 1996 it was apparent that the compressor had a limited life, at that time estimated to be about four years. This limited life was mainly due to the condition of the rotors which had never been replaced. The impellers were of the Z riveted type and with the airborne dirt and occasional pannier intercooler leaks significant corrosion deposits had built up between the impeller covers and the vanes stretching the rivets, see figure 2. The HP rotor condition deteriorated to such an extent that the impellers had to be repaired in 1999 and new vibration trips were fitted in 1998 to detect an in service failure and protect against consequential damage.

An assessment of the machine condition and protection in 1996 had shown that the control system was inadequate for a machine of this age and type and that the pipework and lubrication system were beginning to show their age. This was illustrated in 1998 when a problem with the oil filtration led to a gearbox failure. Fortunately a spare gearbox was available and the machine was brought back online. If the compressor was to continue in service beyond 2001 not only would the LP rotor and intercoolers need to be refurbished but the instrumentation would need to be upgraded and oil filters replaced with modern ones. The T6 PAC reliability was affected both by its condition and that of the systems associated with it and excluding main plant overhauls its availability had averaged 95% for the previous few years. The machine efficiency was also poor compared to the T8 PAC and for the marginal product produced using air from the T6 PAC the electricity consumption was 108% of that required for air from the existing T8 PAC.

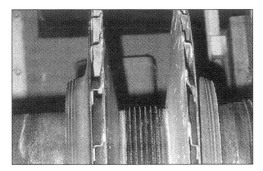

Figure 2 T6 PAC impeller condition

The T8 PAC had been commissioned in 1981 when the plant air requirement had been for 115,000m3/hr at 31bara. Over time plant conditions had changed and the compressor needed to be run flat out at all times and in summer the two compressors were unable to provide enough air. The reduction in delivery pressure to 24bara also meant that the compressor was running significantly off design. In addition to the air compressor the T8 PAC has an expander to recover energy from the offgas produced on the plant. The expander is a fixed volume machine and was too small to swallow all the offgas at the current plant conditions.

2   PROJECT PROCESS

2.1   DuPont Guide to Project Implementation
The process that was used for this project was the DuPont Guide to Project Implementation. This process is split into three discrete steps, Front End Loading, Project Implementation and Facilities Operation. Front End Loading has three stages: Business Planning; Facility Planning and Project Planning. Project Implementation covers detailed design and construction, Commissioning, start-up and project closure by Facilities Operation.

2.2   Project Team
The complexity of the project meant that the project team was flexible and different individuals and organisations were involved at different stages. The project personnel were from five different organisations. The project manager, plant representative, instrument engineer and commissioning manager were from the operator, DuPontSA. The machines engineer, senior process engineer and construction manager from the consultancy arm of the technology provider, DuPont Polyester Technologies. The detailed design contractor, ABB Eutech, provided piping, vessels and instrument designers. Construction resource including planning and supervision was provided by AMEC. The vendor, MAN GHH Borsig, was also part of the project team and their project engineer and technical manager were an integral part of the project implementation team.

3   FRONT END LOADING

3.1   Business Planning
For this project the Business Planning stage was to persuade the business that there was a need to secure a long term source of process air to enable the plant to run at its design capacity. During the previous ten years three projects had been investigated to look at

uprating the T8 PAC and although the financial case was attractive it was felt that the level of capital investment was too great. Following the inspection of the T6 PAC during the plant overhaul in 1999 when it was confirmed that the compressor had a very limited life the business decided to investigate the best method for ensuring a long term reliable source of process air.

3.2   Facility Planning

For this project the major part of the Facility Planning stage was to investigate the options for providing the required process air and then to progress the design to a stage where an initial estimate could be produced to allow the project to be assessed against the business objectives. All of the options also had to be considered against the need to carryout all work requiring a plant shutdown in the identified three week plant overhaul in Q2 2001.

The first option considered was to refurbish the T6 PAC which would require some capital outlay but would bring no variable cost or revenue improvement and the plant would still be rate limited in the summer. This case was used as the base case to compare the other schemes against. Another low risk option considered was to install a new small compressor in parallel to the existing T8 PAC. This would require significant capital outlay but would bring a revenue improvement and a small energy saving. The payback on this option would have been around three and a half years.

Oxygen enrichment has been implemented successfully on other plants and would need a small capital outlay and bring a revenue improvement. Initially this looked an attractive, low risk option but the plant operating conditions meant there would be a significant variable cost penalty and the option was not taken any further. The business were keen to look at any scheme that would minimise the capital outlay so a number of possible suppliers were approached to assess what would be required to supply "air over the fence". The volume and delivery pressure were outside the normal scope of these suppliers and any contract would have had to be for a minimum of ten years and as such this scheme was assessed as too high a commercial risk.

The low risk options considered all had a downside, either the capital cost was high with too long a payback or there was little or no improvement in either variable cost or revenue generation. This led to the team investigating the option of uprating the T8 PAC. This would involve a large capital outlay but would give a revenue improvement and significant energy savings although the project would not be low risk. The payback on this project would be just under two years. The option to uprate the T8 PAC was assessed by the business as most closely meeting the business objectives and the project was sanctioned to proceed to the detailed design phase.

3.3   Project Planning

The first part of the project planning stage was to produce a detailed scope document that could be used to produce an accurate estimate. This scope also enable all the relevant safety and environmental assessments to be completed. At the completion of this stage the project passed the gateway review and full sanction was obtained. The modifications to the machine train covered a new HP compressor casing and rotor but reusing the existing bearing housings, new stator and rotor for the LP compressor and new nozzles for the HP and LP expanders. The machine surge controller and surge limiter were replaced with a modern system along with a number of other obsolete instruments. The first intercooler was found to be too small for the new duty and had to be replaced. The changes to the compressor and the new intercooler also meant that all the process air pipework had to be modified. No changes were required to the gearboxes or the drive motor.

4 PROJECT IMPLEMENTATION

4.1 Detailed design
The focus of the detailed design phase was on reducing risk and eliminating known problems. With relatively short overhaul duration the design had to consider the ease of installation and the dependency of each task. Each individual item was assessed for possible issues and then these issues addressed to minimise any risk posed.

4.1.1 LP compressor casing
The new design for the LP compressor included increasing the number of axial stages from thirteen to fourteen resulting in an increase in the discharge pressure. This meant that under surge conditions the maximum pressure attainable was too high for the rating of the compressor casing. Theoretical calculations showed that the casing design should be adequate but without a hydrostatic pressure test that would prove the casing was capable of withstanding this pressure with a safe margin calculations were not sufficient. The acceptable risk level in DuPontSA for potential Fatal Incidents is 6×10^{-6} per year, without a hydrostatic pressure test the risk for the system was estimated to be 6.4×10^{-5} per year. As a result the possibility of carrying out the required hydrostatic pressure test was investigated. The vendor's previous experience with pressure testing cast iron casings was limited to around 9barg, the requirement for this casing was 12barg. The vendor was sufficiently nervous about a casing failure under hydrostatic test that an alternate method of reducing the risk had to be found. A number of different solutions were investigated but the only one that reduce the risk to a sufficient degree, in this case 1.55×10^{-6} per year, was to install an intermediate relief valve to protect the LP compressor casing.

4.1.2 Rotordynamics
When the existing HP compressor was tested at the vendor's works prior to installation on site the rotor was found to run too close to its second lateral critical speed. The drive end bearing was modified to reduce the second lateral critical speed and bring the operating vibration levels to within acceptable limits. The configuration of the HP rotor is such that the overhung coupling mass gives a second lateral critical speed below the running speed and the expander gives a third lateral critical speed about 30% above the running speed. There have been a number of problems with high vibration throughout the machine's life. The causes have included incorrectly centred coupling shims, the drive end bearing housing sticking on its axial keys, pipework pull on the expander casing and labyrinth rubs. None of these appeared significantly out of tolerance indicating how sensitive the rotor was.

The new rotor design was unlikely to be significantly different from the existing one so it was important that this known problem was considered and every measure taken to prevent a reoccurrence. The design for the new HP compressor has only four radial impellers compared to five for the existing design, the overhung expander impeller is the same as are the bearings and the coupling flange. Overall the weight of the rotors was very similar despite having one less impeller. The rotordynamics for the new rotor were calculated for three different bearing stiffnesses which would result from normal, minimum and maximum bearing clearances. For the normal configuration the calculated lateral critical speeds are:

Order	Lateral Critical Speed	% Running Speed
First	3987 rpm	40%
Second	7803 rpm	78%
Third	14067 rpm	141%

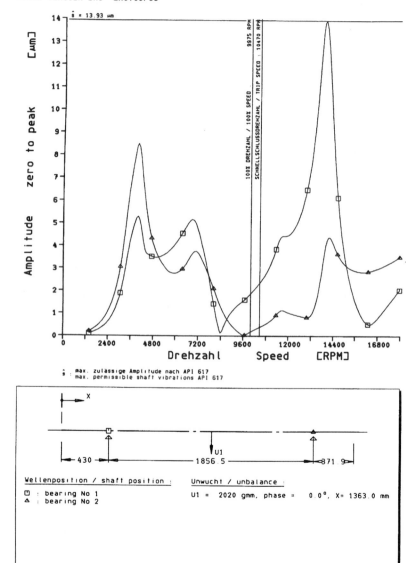

Figure 3 Lateral critical speed predictions for the HP rotor

C603/011/2003 © IMechE 2003

The predicted vibration amplitudes are shown in figure 3. The separation of the second and third critical speeds is better than that for the existing rotor but it will still be sensitive to unbalance. To ensure a sufficiently precise balance, a high speed balance is always required.

4.1.3 Intercooler
Checks were made on the available heat transfer area of the intercoolers to see whether they could be reused. The first intercooler was found to be undersized for the existing duty and had to be replaced as part of the project. The original intercoolers had water on the carbon steel shell side and titanium tubes to prevent corrosion on the air side. There was not enough space for a larger replacement intercooler of the same design so an alternative had to be found. The existing intercoolers have suffered from fouling on the cooling water side and with the water being on the shell side cleaning has been almost impossible. To improve this the new intercooler had to be designed with the cooling water on the tube side. Intercoolers of this design are now the norm in industry but the outlet temperature of 225degC from the axial compressor was too high to allow the use of a coated carbon steel shell and stainless steel tubes. To cater for the temperature the shell of the new intercooler was made from 316L stainless steel and the tubes, tubesheet and return channel were made from Duplex stainless steel to prevent stress corrosion cracking as a result of chloride in the cooling water. One other significant issue was the thermal expansion of the intercooler. Originally a single pass on the air side was used but the shell would have bent by as much as 30mm so a two pass design was utilised to reduce the distortion.

4.1.4 Pipework and layout
The compressor is skid mounted and all the pipework had been fabricated and installed prior to the original installation on site. The arrangement is very compact and modifying the pipework was a major challenge. The first requirement was to get accurate measurements of the existing pipework before any new drawings could be produced. Plant Applications and Systems for Concurrent Engineering (PASCE) was used to get a three dimensional model of the compressor and all its associated pipework and vessels. Once this was available the new layout including the method for removal of the old intercooler and installing the new one could be worked out. The PASCE system was also used to check the pipe stresses and nozzle loads as well as to produce the isometrics for the new pipework.

4.1.5 HP compressor casing
The new compressor casing was larger than the old one so ensuring that it would fit into the existing supports was vital. At the plant overhaul in 1999 detailed measurements of the old casing were taken by the vendor to check against the original drawings to allow the new design to take any differences into account. The PASCE model also provided information to check the new casing dimensions.

4.1.6 Expander
To increase the capacity of the expander new nozzles had to be fitted into both the HP and LP casings. Previous experience has shown that getting the blades to fit accurately into a 360 degree ring can be difficult. To reduce the risk of a fitting problem the manufacturing tolerance on the blades was considered and a number of slightly different sized closing blades were provided and the blades were test assembled into a dummy ring to check the fit.

4.1.7 Instrumentation

The changes to the machine instrumentation were fairly significant and it was realised that there would need to be a lengthy testing procedure prior to commissioning. To reduce this a detailed factory acceptance test was carried out by the vendor and the DuPontSA instrument engineer prior to shipment of the surge controller so that the on-site testing scope would be reduced.

4.1.8 Schedule

Initially a plan was produced for the machine installation work with key hold points for installation of pipework and instruments. The instrument installation plan was then integrated into the overall plan and finally the intercooler and pipework installation added. This composite plan was then approved by all parties involved and formed the basis for resource allocation. An outline of the overall schedule is shown in figure 4, the detailed plan had over five hundred entries encompassing all the work in the compressor house area and is on a 24 hour basis to meet the required timescale.

ID	Task Name	Duration
1	**Project Duration**	**540 hrs**
2	Train Isolation	12 hrs
3	**LP Compressor and Expander**	**264 hrs**
4	Open casings and remove rotor	36 hrs
5	Check bearings	4 hrs
6	Reblade expander	96 hrs
7	Modify compressor casing	96 hrs
8	Fit new compressor internals	72 hrs
9	Fit rotor and check alignment	48 hrs
10	Fit top covers	12 hrs
11	**HP Compressor and Expander**	**228 hrs**
12	Open casings and remove rotor	36 hrs
13	Check bearings	4 hrs
14	Reblade expander	72 hrs
15	Remove old casing	24 hrs
16	Fit new bottom half casing	48 hrs
17	Fit new rotor and check alignment	96 hrs
18	Fit top covers	24 hrs
19	**Instruments**	**384 hrs**
20	Fit new field instruments	120 hrs
21	Fit Turbolog	48 hrs
22	Connect all instruments	144 hrs
23	Test	72 hrs
24	**Intercooler and Pipework**	**336 hrs**
25	Disconnect and remove redundant pipework	24 hrs
26	Remove old interooler	48 hrs
27	Fit new intercooler	72 hrs
28	Fabricate new pipework	144 hrs
29	Align pipework to compressor casings	48 hrs
30	**Train**	**72 hrs**
31	Alignment	24 hrs
32	Oil Flush	48 hrs
33	**Commissioning**	**120 hrs**
34	Turbolog acceptance test	24 hrs
35	Trip testing	24 hrs
36	Test Run	72 hrs

Figure 4 Outline Schedule

4.2   Construction

4.2.1   Preparation

To allow those working on the construction to be familiar with the work required a dedicated team be assembled in advance of the overhaul. This integrated team included labour for the execution phase, management and supervision and a dedicated technical resource. The team was led by a construction manager who had responsibility for all disciplines. There was a duplicate team on the night shift. The detailed plan was used to predict and resource the labour required in advance.

4.2.2 Execution

The level of planning and detailed design for this project meant that very few major technical problems were encountered during the execution phase. As expected the compact design of the installation did lead to a number of problems. The space for removing the old intercooler and installing the new one was very tight and this proved a challenge for the rigging team. The coolers had to be manoeuvred on skates because there was no direct overhead access for the cranes. Despite the new intercooler looking like it was larger than the space available, the use of the PASCE system meant that the team was confident it would fit, see figure 5. When the intercooler was installed the clearance on each side was around 50mm with 10mm clearance above to the main oil return pipework see figure 6. Installing the intercooler proved

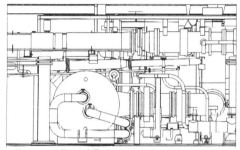 

Figure 5 PASCE drawing for intercooler   **Figure 6 Intercooler in position**

to be relatively straightforward compared to the pipework modifications. Most of the old pipework could not be removed without cutting it into sections so most of the pipework was to be modified rather than totally replaced. As expected lining up the old and new sections of pipe was not always easy and getting the set up right before welding was time consuming although to the credit of the fabrication team none of the major pipe welds had to be repeated. Aligning the pipework to the machine was one of the critical tasks for the execution phase. It took time to get the pipe supports set up correctly but once this had been done all the pipe to nozzle alignments were within tolerance and accepted by the vendor.

Figure 7 LP compressor internals   **Figure 8 HP compressor casing**

During the 1999 overhaul accurate measurements had been made of the HP compressor but the LP compressor had not been opened so it had not been possible to determine the extent of any machining to fit the new internals. It had been hoped to do any modifications by hand but was fairly slow but once in place the modifications were made fairly quickly and there were no further problems installing the new internals see figure 7. The installation of the HP compressor went according to plan with only minor modifications required see figure 8.

The other major modifications were around the machine control system and the new instrumentation. Installing the instrumentation went smoothly but the final set-up and testing had to be delayed until the oil flush had been completed. There was limited instrument resource and often no night shift cover for the final testing and this took longer than expected. In retrospect it would have probably been easier to install a whole new system rather than try and integrate the new surge controller into the existing instrumentation.

5   FACILITIES OPERATION

5.1   Commissioning and testing
The commissioning of the machine system had been planned in the same detail as the execution phase of the project. A detailed risk assessment had been carried out and operating instructions with associated checksheets for all operations had been produced in advance to ensure that all personnel knew what was required. Unfortunately at the first machine start it was immediately apparent that there was a problem with high vibration on the HP compressor. However the signals for the vibration probes had been recorded and the initial indication was a rub on the HP rotor. This was rectified but a further start showed that the rub was not the root cause. Finally the cause was traced to the coupling bolt holes having been drilled 0.07mm off centre. It was not possible to correct this directly so a number of lighter coupling bolts were used to correct the unbalance load that was being caused by the incorrectly drilled holes. Once the coupling unbalance had been corrected the rest of the commissioning program went to plan.

After the successful completion of the API mechanical test the compressor was handed over to the operation team for the plant to be brought on-line. Once the plant was operating consistently a compressor performance test was carried out. This showed that the compressor was capable of delivering the head and flow required but that the power consumption of the HP compressor was greater than predicted.

5.2   Outstanding issues
The main outstanding issue following commissioning is the greater than expected power consumption of the HP compressor. Investigations are in hand with the vendor and a number of possible causes have been identified. The project team is confident that the power consumption can be improved to meet the original guarantee without major modifications.

6   SUMMARY

This project has shown that it is possible to successfully carry out a major machine uprate in a short duration overhaul. Extensive planning and preparation is required to minimise the number of unexpected problems but the project team needs to be flexible enough to handle any surprises when they occur.

C603/020/2003

Technical specification of turbomachinery rerates

S A RODGERS and **C J ROBINSON**
PCA Engineers Limited, Lincoln, UK
J O'CONNOR
Hickham Industries Inc., Houston, Texas, USA

ABSTRACT

Performance 'rerating' of existing turbomachinery typically involves incrementing the power from a turbine or adjusting the flow-speed relationship of a compressor to reflect changes in plant requirements. Rerating can be more cost-effective than the purchase of new equipment, particularly when new hardware is already required during a scheduled machine overhaul. The technical challenge is to achieve the desired performance with minimal changes to the hardware, and certainly within the existing casings. Some of the methods used to determine the necessary hardware modifications are illustrated via the detail of a steam turbine rerate.

1 INTRODUCTION

One segment of the turbomachinery market that has seen significant growth over recent years is the so-called performance 'rerating' of existing hardware. Worldwide, industries employing turbomachinery are adapting as markets change, expand and become evermore competitive. Rather than suffer the delays and expense of purchasing and installing new turbomachinery, an increasing number of operators are choosing to make modifications to their existing hardware during scheduled downtime for repair and overhaul.

There are turbomachines of all shapes and sizes working in many different industries around the world. Consequently, there are diverse reasons for rerating. In general, however, most rerate projects involve incrementing the power from a turbine or adjusting the flow-speed relationship of a compressor to reflect changes in plant requirements. In many such cases, it is simply not cost-effective to purchase new turbomachinery when appropriate performance-improving modifications may be made to the existing machine during scheduled downtime.

Steam turbines, for example, usually have a lifespan much longer than the equipment around them. Hence, it is not uncommon for the duty of a steam turbine to change as elements of the surrounding plant are replaced and improved over time. So long as these plant changes do not constitute a significant departure from the original design intent of the turbomachinery, a number of relatively inexpensive modifications to the blading may be all that is required to meet the new plant conditions.

Over the years, as the proportion of 'old' turbomachinery in operation has increased and rerating has grown in acceptance and popularity, turbomachinery manufacturers, overhaul specialists and professional engineers have had to adapt traditional design methods to solve new sets of problems. Indeed, with each new project 'rerate engineers' have to develop a familiarity with the details of plant operations in order to identify the rerate targets. This adds a layer of complexity to what might otherwise be a relatively simple overhaul.

Once the potential is identified, it is necessary to ascertain that the rerate targets are technically realisable before quoting for the work. To mitigate risk, it is highly desirable for the operator to invest in an engineering study that thoroughly investigates the feasibility of the rerate demands. Having made this contribution, the operator may then receive the benefits of tighter tolerances on performance guarantees.

To keep the price attractive to operators, as well as for practical and logistical reasons, the challenge is to limit the scope of potential modifications and effect the desired change in performance with minimal changes to the hardware, and certainly within the existing casings. Constrained to the existing gas-path, and with the upstream and downstream boundary conditions prescribed by the plant requirements, the scope of potential modifications is often limited to the blading. Even with this constraint, it is likely that a number of rerate options can achieve the indicated performance. Indeed, the iterative process of developing such rerate options can identify important performance measures and unexpected patterns of behaviour, so it is desirable to consider as many options as possible in the time available. Still, the cost of unscheduled or extended machine downtime can exceed the cost of an overhaul, so it is imperative that the optimum rerate option is chosen in good time for the manufacture of any new hardware. For these reasons, fast and accurate computational tools are of paramount importance at this stage in the process.

All rerate projects begin with an analysis of the existing machine, which the operator may be unwilling or unable to take offline for detailed measurement. Sometimes an operator will have complete sets of drawings and accurate performance data but often the rerate engineer has the significant technical challenge of distilling an adequate understanding of the performance from an incomplete description of the machine geometry and sparse measurement of performance data.

The usual technical strategy employed is to synthesise the performance of the existing machine using 1D and 2D semi-empirical methods – embodied as 'meanline' and 'throughflow' computer codes. These computational tools include correlations to enable performance to be predicted from a fairly high-level knowledge of the hardware, rather than the detailed information such as the coordinates of blade profiles. They are sufficiently accurate in their capturing of the physics that the unknown aspects of the geometry can be guessed and iteratively adjusted until the predicted performance matches the measured data.

Using this synthesis performance as a baseline, the hardware adjustments necessary to achieve the desired incremental performance can be derived.

The rerate design process is illustrated below via the detail of a steam turbine rerate performed by PCA Engineers Limited and Hickham Industries Inc. in 1999. Also included is a summary description of the meanline and throughflow codes used to complete the project.

2   DESCRIPTIONOF THE CODES USED

Since computers have been available, 1D 'meanline' and 2D 'throughflow' codes have been used to aid the design of turbomachinery. These codes solve the equations of energy, continuity and, in the case of throughflow codes, radial equilibrium, using empirical correlations to predict the exit gas angles and pressure losses for many standard blade profiles. Other losses, such as those due to boundary layer growth, lacing wires and blade tip-clearances, can also be modelled, again using semi-empirical methods resulting from extensive testing of different types of blades.

In a rerate analysis, meanline and throughflow models of the existing hardware may only be achieved after a number of blade and other loss models are investigated, particularly when the machine to be rerated is old, damaged or otherwise unique. Having satisfactorily replicated the baseline performance, a large number of modified hardware configurations may then be simulated before the desired levels of performance are achieved. A rerate analysis is thus highly iterative and great demands are made on the flexibility of the loss models included in the meanline and throughflow codes used.

Today, fully three-dimensional (CFD) codes solving the Reynolds averaged Navier-Stokes equations are widely available, relatively inexpensive, and can, at least in principle, be used to design turbomachinery components. However, the use of CFD for the direct design of turbomachines remains limited, as the calculation times far exceed those of typical throughflow codes, particularly for multi-stage machines. In addition to the time overheads, the major obstacle to widespread use of CFD for rerate analyses is simply a lack of adequately detailed geometry. At PCA Engineers, CFD is used mainly to underwrite key technical changes close to the end of the rerate process, when a more accurate picture of the geometry has been formed.

The rerate project outlined below was completed using meanline and throughflow codes developed over many years by PCA Engineers Limited.

3   RERATE OF A STEAM TURBINE

In general, steam turbines are not as straightforward to rerate as gas turbines. Curtis stages, two-phase flows, extraction and admission valves, partial admission nozzles and inter-stage plenums are just some of the features of steam turbines that demand special treatment. Methods for handling a number of these features are described below.

In 1999, PCA Engineers was contracted by Hickham Industries Inc. to determine the rerate potential of a ten-stage De Laval steam turbine, producing 4991 kW from 25793 kg/hr of steam. The operator originally requested a modest power increment of 268 kW (5%) but later revised this figure to the maximum possible increase in output. Due to changes to machinery elsewhere in the plant, the operator also required that the rerated turbine run at an increased speed of 5100 rpm, up from 4860 rpm. The baseline turbine is illustrated in figure 1 and in cross-section in figure 2.

Rotor assembly   Diaphrams

Figure 1 – Stripped-down, baseline turbine

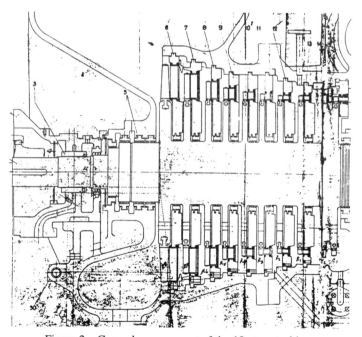

Figure 2 – General arrangement of the 10-stage turbine

Figures 1 and 2 are typical illustrations of the state of the hardware approaching a rerate exercise and the quality of the available documentation.

3.1 Computational Modelling of the Baseline Hardware

As with all rerate studies, the first task was to construct a computational model of the existing hardware from measured geometry and boundary conditions and then adjust the model to calibrate it with the available performance data. In this case, the model had to be developed from field measurements of the turbine. This model was then progressively refined as components became available for more precise measurements with a coordinate measuring machine (CMM). The condition and location of this turbine made field measurements problematic and the field measurements of many components were found to differ quite significantly from those made using the CMM. Only the results for the most accurate set of measurements are reported here.

The first section of the turbine was of Curtis configuration, with a partial admission inlet nozzle, two sets of rotor blades and a set of closely spaced reversing channels. Only 37.5% of the channels in this first nozzle were open to admit the flow of steam. For convenience, the two Curtis rows will hereafter be referred to as C1 and C2 and the latter stages will be referred to as stages R1 to R8. The R1 nozzle also provided partial admission, through approximately 89% of the channels.

Table 1 outlines the performance of the turbine before the rerate:

Inlet stagnation pressure	(kPa)	2862
Inlet stagnation temperature	(K)	644
Steam flowrate	(kg/hr)	25794
Outlet pressure	(kPa)	13.5
Rated power	(kW)	4991.0
Rotational speed	(rpm)	4860

Table 1 – Performance of the baseline hardware

Preliminary calculations using the above data and a set of steam tables determined the state of the steam at the turbine exit and the thermal efficiency of the turbine. The expansion of the steam from a superheated state at inlet to a saturated condition at the turbine exit is plotted in figure 3. The information suggested a steam quality of around 94.8% at turbine exit and thermal efficiency of around 72.4%.

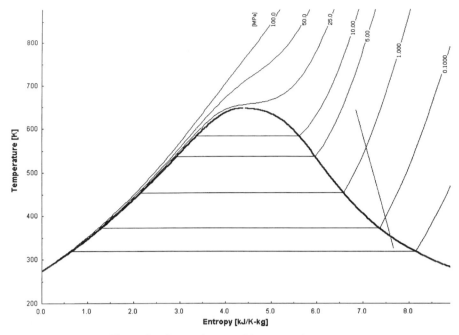

Figure 3 – Approximate turbine expansion curve

The meanline and throughflow codes PCA uses to analyse and rerate gas and steam turbines, TPM and SC90T, assume that the turbine annulus is represented by axially spaced calculating planes intersecting the hub and shroud, including planes coincident with all blade leading and trailing edges. The programs work in a similar way to solve the flow equations at each axial plane, with TPM solving at only the root-mean-square (RMS) radius and SC90T at a number of radial stations distributed from hub to shroud. To compute blade losses and exit flow angles, the codes require geometry for each blade at a number of spanwise positions. The exact blade profiles are not required; instead the correlations need only the blade angles and certain ratios, like pitch/chord and throat/pitch. The necessary blade data were quickly derived from the CMM measurements of the blades.

Because of its configuration, the ten stages of the turbine could not all be included in a single computational model. Deep inter-stage plenums made it convenient to model the turbine in four sections, which were then matched based on the swallowing capacities predicted by SC90T, appropriate corrections being made for the effects of partial admission, wetness and plenum losses on flow capacities and efficiency. The turbine was separated into the following sub-models:

1) C1 and C2 modelled as full admission stages
2) R1 modelled as a full admission stage
3) R2 to R4
4) R5 to R8

In 1999, steam properties had not yet been incorporated into PCA's meanline and throughflow codes, so the superheated steam was reasonably assumed to be an ideal gas. For most working fluids TPM and SC90T will calculate specific heat at each axial plane from the local temperature, but because of the transition to saturated steam, where the usual definition of specific heat is invalid, it was necessary to specify for each sub-model a constant, mean value of specific heat derived from graphs of the specific heat of steam as a function of temperature and pressure [1]. Wetness was taken into account using Baumann's rule [4]: debiting the predicted efficiency by 1% point for every 1% of wetness.

Losses due to partial admission were accounted for following a description of impulse stages given in [2]. Figure 4 was generated from this description and used to define an additional efficiency debit for each of stages C1, C2 and R1, which were most strongly affected by partial admission.

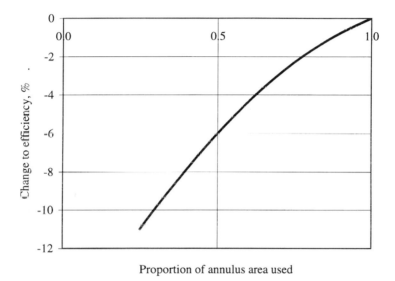

Proportion of annulus area used

Figure 4 – Efficiency debit for partial admission stages

Both TPM and SC90T provide the user with the flexibility to increase or decrease the calculated loss coefficients to better model old or damaged blades and other loss generating features in the gas-path, so plenum loss coefficients derived from [3] were added to the calculated blade loss coefficients (Y).

At the first nozzle in each sub-model:

$$Y_{Pl,N} = \frac{1}{2} \frac{\rho_1 V_{abs,1}^2}{\rho_2 V_{abs,2}^2}$$

At the last rotor in each sub-model:

$$Y_{Pl,R} = \frac{\rho_2 V_{abs,2}^2}{\rho_2 V_{rel,2}^2}$$

where: ρ = density
V = velocity
$_1$ = leading edge
$_2$ = trailing edge
$_N$ = nozzle
$_R$ = rotor

The matching of the four turbine sections was achieved by generating characteristics of corrected mass flowrate, $\dot{m}^*$, and corrected power, $\dot{W}^*$, for each section over a range of pressure ratios.

$$\dot{W}^* = \frac{\dot{W}}{P_{0,1}\sqrt{T_{0,1}}} \qquad \text{and} \qquad \dot{m}^* = \frac{\dot{m}\sqrt{T_{0,1}}}{P_{0,1}}$$

Having established the performance of each sub-model, the performance of the ten-stage turbine was predicted using an iterative procedure based on the flow capacity of stage C1 and the total pressure across the turbine. The powers produced by stages C1, C2 and R1 were determined by assuming pressure ratios and then factoring the results for the full admission sub-models by the proportion of the nozzles admitting flow and by a ratio of efficiencies:

$$\frac{\dot{W}_X}{\dot{W}} = \frac{A_X}{A}\frac{\eta_X}{\eta}$$

where: W = power
A = nozzle area admitting flow
η = section efficiency
$_X$ = partial admission section

The contributions from the remaining sections of the turbine were then calculated by working downstream through the turbine. The pressure ratios across the sections were then iteratively revised until the correct flow capacities and turbine exit pressure were achieved.

3.2 Computational Modelling of the Rerate Configuration

If the overall efficiency and stage loadings remain the same, the relationship between the power produced by a turbine and the mass flowrate is perfectly linear. Relating this ideal behaviour to a rerate design, a 10% increment to the power output may be achieved by making modifications providing a 10% increase to the equivalent throat area of the turbine. So long as the first nozzle is not choked, the equivalent area of a turbine may be calculated using:

$$\frac{1}{(equivalent\ area)^2} = \sum_{i}^{n}\left(\frac{1}{N_i^2} + \frac{1}{R_i^2}\right)$$

where: N_i = throat area of nozzle i

R_i = throat area of rotor i

n = number of stages

In practise, however, it is not generally possible to increment the power by increasing the throat areas of all blade rows equally because of inevitable changes to the stage pressure ratios and loadings, neither is it desirable. To make projects economically viable, it is always the aim to effect the desired change in performance with the minimum changes to the hardware.

Regardless of the number of blade rows that are modified to achieve the uprate, it is generally expected that the operator is able to provide more steam. Incrementing the power of a condensing steam turbine by increasing its efficiency can be extremely problematic, simply because of the increase in wetness and rates of blade erosion in the LP stages of the turbine. In this case, the operator was able to provide as much steam as the turbine could be made to swallow.

A number of rerate options (Options A to J) were developed by simulating modifications to both nozzles and rotors, all giving approximately the same equivalent throat area. All options retained the original 37.5% partial admission for the Curtis stages, except Option II which investigated the benefits of increasing this proportion of the annulus to 48.6%. In the event, there was doubt about the practicality of using a greater number of channels in the C1 nozzle without changing the steam chest castings, so Option H was discounted. All options involved rotating, or 'restaggering', the blades to increase the available throat area. Predicted powers for several of these options are plotted in figure 5, below.

The rerate design recommended to the operator was Option J, the third point from the right in figure 5, which involved raising the shroud radius, lowering the hub radius and opening the throats of the nozzles in the two Curtis stages. Stage R1 was made a full admission stage; the R1 nozzle was restaggered by 2.7°, and the remaining nozzles were restaggered by 3.5 to 5.1°. No rotor profiles were modified.

Although predicted to produce more power, Option I was felt to be too costly as it involved modifying the two Curtis rotors. Wherever possible, it is best to modify only the stationary components of a turbine, thereby avoiding the unnecessary transport of the rotor and the necessity of ensuring the dynamic integrity of new hardware.

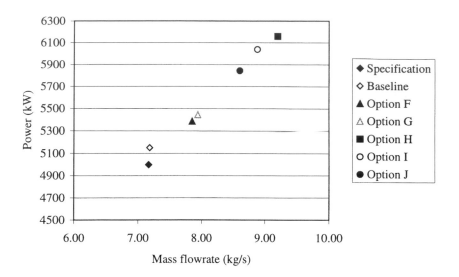

Figure 5 – Power vs. massflow

The importance of the inlet piping and valving was discovered quite early in the analysis, and it was concluded that the rerated turbine required a properly reconfigured inlet pipe system to ensure that the increased mass flowrate did not also result in a decreased turbine inlet pressure. Analyses of the main valve, inlet chest, control valve and steam chest indicated that the pressure drop to the turbine inlet would increase from 560 kPa to nearly 920 kPa with an additional 30% steam flow unless changes could be made to these components.

Of course, opening the angles of blades in a turbine changes the stage reactions and affects the axial thrust. Although many older steam turbines have thrust bearings that are heavily over-engineered and can easily support the additional loads resulting from rerate modifications, it is generally necessary to verify that the changes in aerodynamic thrust are not excessive. Finally, Hickham Industries Inc. used a readily available Finite Element software package to check that the components of the rerated turbine were not subjected to resonant frequencies or excessive stresses.

3.3 Performance of the Rerate Configuration
Table 2 outlines the predicted and actual performance of the turbine in the recommended rerate configuration on a day in Spring 2002. As sometimes happens, the operator had chosen not to increase the speed of the train; in fact, on the day the performance measurements were made, the steam turbine was operating at decreased speed relative to the original configuration:

                                   C603/020/2003 © With Author 2003

		Baseline	Predicted (1999)	Actual (2002)
Inlet stagnation pressure	(kPa)	2862	2862	2963
Inlet stagnation temperature	(K)	644	644	639
Steam flowrate	(kg/hr)	25794	30960	30450
Steam flowrate increase	(%)	–	20.03	18.05
Outlet pressure	(kPa)	13.5	13.5	~10.4
Power	(kW)	4991.0	5842.7	~5874.1
Power increase	(%)	–	17.06	~17.69
Rotational speed	(rpm)	4860	5100	4573

Table 2 – Performance of the rerated hardware

Now, it is not uncommon for rerate projects to be undertaken in locations that are relatively inaccessible; indeed, turbomachinery rerates are probably most attractive to plant operators in geographically remote regions, great distances from the bases of the original equipment manufacturers (OEMs). This steam turbine is just such an example. The turbine is located in provincial Colombia, in a region renowned for its terrorist activity, and it has been difficult to maintain communications with the operator and obtain accurate performance measurements. Although there is some uncertainty over the actual exhaust conditions presented in Table 2, the operator appears satisfied with the performance of the turbine.

From a purely economic viewpoint, this rerate project has been a great success. To rerate the steam turbine and the upstream air compressor (i.e. the entire train), along with delivery to site of the new internals for the steam turbine, the overhaul of both turbomachines, labour and supervision and start-up commissioning, cost the operator US$1.1m. Excluding the time for the aerodynamic and mechanical engineering studies outlined above, all work was completed within 6 months.

By comparison, replacing the steam turbine and the air compressor with new hardware would have cost an estimated US$4.7m. Delivery, installation and commissioning might have taken as long as 18 months from receipt of order. Alone, a new 5.2 MW steam turbine would have cost the operator US$950,000, exclusive of agents' fees, labour and supervision costs, and the costs of civil, mechanical and instrumentation engineering.

4   CONCLUSION

Performance rerating of existing turbomachinery has grown in acceptance and popularity over recent years. The most typical rerate projects involve incrementing the power output from a turbine or adjusting the flow-speed relationship of a compressor to reflect changes in plant requirements, often during scheduled downtime for repair and overhaul. If the desired levels of performance can be achieved with minimal changes to the hardware, rerating can be an inexpensive alternative to the purchase of new machinery.

Turbomachinery rerate projects are often subject to inflexible time constraints and the detailed geometric information required by CFD codes is usually not available. It can be a significant technical challenge to distil an adequate understanding of the performance of a machine when the operator is unwilling or unable to take it off-line for detailed measurement. Capable of predicting performance accurately from a relatively high-level knowledge of the hardware and flexible enough to allow modified configurations to be analysed quickly and easily, meanline and throughflow codes are ideally suited to rerate projects. Despite the rapid development of CFD, codes performing calculations in 1D and 2D remain indispensable tools for the design and analysis of axial flow turbomachinery.

This report has attempted to illustrate the level of engineering analysis that supports typical rerate projects via the detail of a steam turbine rerate. Recent measurements from the rerated turbine compare favourably with computational performance predictions obtained from meanline and throughflow codes. The overall costs of rerating the existing steam turbine train have been weighed against the estimated costs of purchasing and installing new hardware in a difficult location.

REFERENCES

[1]   J H Keenan, F G Heyes, P G Hill and J G Moore,
       Steam Tables (SI units),
       John Wiley & Sons, 1969

[2]   K Korematsu and N Hirayama,
       Performance estimation of partial admission turbines,
       ASME 79-GT-123

[3]   D S Miller,
       Internal Flow Systems, 2nd Edition,
       BHR Group Ltd, 1996

[4]   J B Young,
       Wet Steam Flows,
       University of Cambridge Programme for Industry,
       Turbomachinery Aerodynamics, June 1996

[5]   D Flaxington & P M Came,
       Throughflow Analysis of a ten-stage DeLaval Steam Turbine,
       PCA Engineers Limited, May 1999

C603/027/2003

Uprate applied to turbomachinery in olefin plants from MAN Turbo's point of view

G HINRICHS
MAN Turbomaschinen AG (MAN TURBO), Germany

MAN TURBO'S MACHINERY IN EUROPEAN OLEFIN PLANTS

In the last 35 years, more and more ethylene plants have been erected in Europe MAN Turbomachinen AG, formerly Gutehoffungshütte Sterkrade, has delivered more than 100 casings to 30 different ethylene plants in Europe

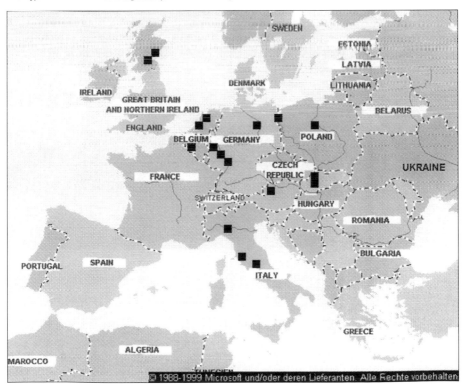

In the course of their operating time many plants have been modified to meet market requirements. Modifications have been made to change in feed stock and capacity, environmental protection (statutes relating to emissions), efficiency and reliability.

The diagram shows the number of new installations and uprates and in addition the uprates broken down into the main reasons for conversion.

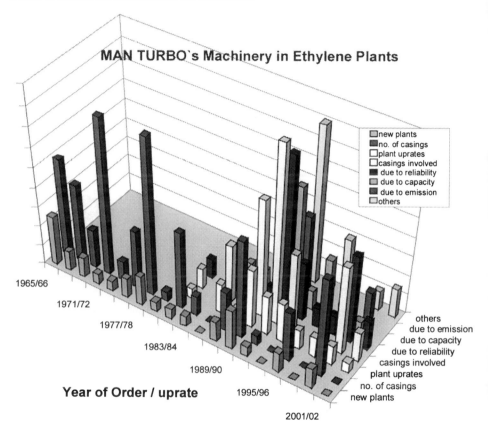

In the past , the largest number of conversions/modifications were implemented in response to changed market situations and/or adaptation to changed feedstock parameters such as pressure, temperature and chemical composition.

The target was to increase production within geometric limits if possible; increased speed brought about by modifications to gear units (new gear sets etc.); and/or utilization of speed reserves in the turbine and compressor was one of the messures.

Modern calculation methods (finite elements) on one hand allow a more precise observation of the maximum possible loads of the dynamic components. The application of a modern electronic overspeed trip system which enables not only the absolute value of the speed as an switch-off criterion to be drawn upon (as in the case of a mechanical trip release) but also includes the vector of the speed change on the other hand, increases and secures existing reserves. Increases in capacitiy of up to 10 % are thus possible.

Changing or modifications of assembled components and rotors have been an other solution to uprate machines. Due to the use of impellers with a higher absorption capacity and levels of strength as well and corresponding adaptation of the internal components (volutes), capacity increases of up to 30 % have been achieved. Even on occasion, replacing a entire compressor casing or the turbine in a train has been a measure to increase the capacity of the plant.

Although uprates and operating changes have already been implemented in many plants, there still remains much potential that can be tapped. As an example, in 2002 only, studies for expanding three plants with a total of 15 casings were drawn up.

Optimizing the plant for highest availability at maximum usage, an application-specific profile must be drawn up for every plant component. This implies meeting the demands of implementing the most recent technical standards as well as fulfilling environmental protection standards.

I would like to look more closely to two issues arising from the wider field of opportunities for optimization:
- Measures to prevent fouling or to minimize the negative effects of fouling, and
- Optimizing the shaft sealing system to increase availability by converting oil stuffing boxes (glands) and/or mechanical shaft seals to gas seals.

Compressor fouling
Compressor fouling has always been an accepted fact in ethylene manufacturing. Attempts to minimize fouling have not been very successful.

Compressor fouling occurs, when during the course of the operating time at a constant operating pressure, there is a decrease in volume concurrent with an increase in the discharge temperature. This relationship is brought about by deposits of polymerizates in the flow area of the compressor

- Deposits of polymerizates on volute and diffuser -

What causes fouling?
First of all the feedstock. In an untreated gas, the content of hydrocarbons of a higher valency in addition to constituents such as sulphur compounds, nitric oxide and peroxide are primarily responsible for the susceptibility of the gas to polymerization.

Secondly temperature: The discharge temperature of the stage groups depends on the number of intercooling stages (normaly 4) and the pressure ratio.
 - These parameters are defined in the process specification.

At the compressor design point, the discharge temperature is always below 100°C. However, the polymerization of untreated gases commences much earlier and the rate of polymerization increases rapidly at temperatures greater than 85°C.

Third is ressure, because fouling increases almost proportionally to pressure and at last the surface finish. As one can imagine, a rough surfaces are prone to deposits, smooth surfaces are less prone to fouling

The consequences of polymerizate deposits in the compressor for the operator with respect to the efficiency rating and availability are at first a decrease of production.

At a given speed the discharge head/volume flow characteristic is lower and the efficiency curve tails off accordingly. Increasing the speed can counteract this, however, the original volume is now achieved at a less favourable efficiency level and at a higher discharge temperature. The speed of polymerization increases and the fouling process accelerates as a result.

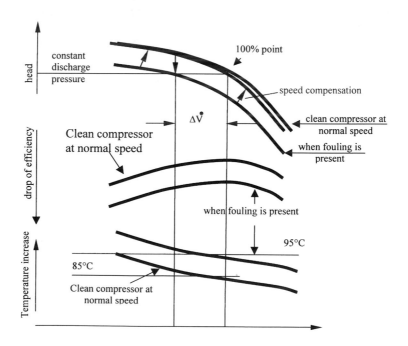

C603/027/2003 © With Author 2003

At second the mechanical behaviour of the compressor changes due to unbalance. Up to a certain point, deposits are evenly distributed and are not registered by the shaft vibration monitoring system. It is only later that uneven build-up of deposits and partial cracking of the coatings on the rotor lead to unbalance, high shaft vibrations and eventual shutdown.

At last wear of parts can be expected. When the flow passages and the spaces between the rotor and the stationary parts become constricted this in the long term my lead to mechanical damage. The rotating parts are damaged by an emery (abrasive) effect. The axial thrust of the rotor changes and the thrust bearings can get overloaded, leading to a turbine trip. Unequal impeller flow channel geometry can lead to a stage surge and this in turn can cause vibration-induced fractures on the shroud discs of the impellers.

So the question is, how fouling can be counteracted effectively? Preventing or minimizing the formation of polymerizates on components in the flow pathes is the answer. It can be done by developing smooth surfaces, avoiding dead areas and narrow passages and apply in addition a non-stick coating such as Sermalon to the flow passages of stator and rotor.

Reducing the discharge temperature at every stage by constant injection of water up to the saturation limit a reduction of up to approx. 10°C can be achieved. The cooling process results from the evaporation heat loss (approx. 90%) and the rise in the water temperature (approx. 10%). This has in addition the positive effect that the requisite compression work isreduced by the cooling process. However, because the total molecular weight is reduced and the mass flow is increased by the corresponding proportion of water vapour, the advantage is somewhat reduced.

By injecting water in cracked gas compressors some prerequisites have to be followed:
- Ascertain the volume exactly by using calculation programs / water vapour tables,
  Water should have a pH value between 8 and 9.5,
  The water/gas differential pressure at the injection point should be approx. 8 bar,
- The average droplet size should be less than 100 microns,
- Injection should take place in the gas path after each impeller. Ideally are at least 2 nozzles installed at every change of flow direction.

Washing/cleaning the surfaces by constant injection of extra water above the saturation line in order to achieve a washing effect is an other measure.

Sporadic injection with condensate, washing oil etc. was the practice in the past and still is in many cases. It is not recommended by MAN Turbo. The removal of the deposits is rather uneven; the original condition cannot be reinstated by this means. Nozzles clog up lightly during the periods between washing. Deposits may be removed to some extent out of the compressor, however, they are then re-deposited more easily in the columns/heat exchangers downstream.

Example of a revamp of a 30 years old feed gas compressor from oil lubricated seals to dry gas seals
Since the compressor was first commissioned, the sealing system caused a lot of problems:
- Very high maintenance requirements caused by product deposits in process pipes and in the oil system,
- Change of seal oil at relatively short intervals and emissions caused by required seal oil disposal and discharge of oil vapour over the roof.

- Service life was limited, because overhaul interval has been reduced to 1-2 years due to clogging and oil penetration into the product.

A) Seal oil supply
B) Degassing line to vent
C) Oil / gas mixture to Flowtrap
D) Oil return

**old
oil bushing
design**

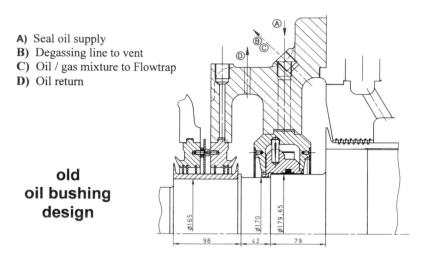

By changing over to a gasseal design, the old seal housing was replaced by a modified design meeting the requirements of the machine (accommodation of the casing guide) as well as those of the seal.

A FLOWSERVE duplex dry gas seal type GASPAC 987 was installed. The sealing of the bearing housing was integrated into the sealing concept by installing an ESPEY carbon ring seal buffered with nitrogene.

A) Buffergas supply
B) Degassing line to vent
C) Leakage gas seal (atm. side)
    and buffergas leakage from port D
D) Buffergas supply bearing housing seal
E) Cracked gas / N2 - mixture

**new
gas seal**

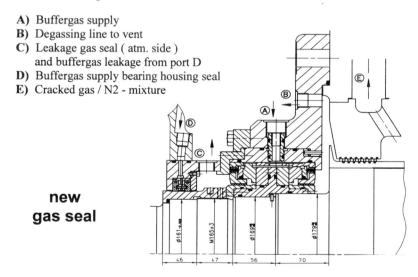

C603/027/2003 © With Author 2003

The control panel was devided into
- A central buffer gas supply for two trains. The main supply is provided via the 69 bar bus with safeguarding via a backup in the event of gas supply failure and an additional safeguarding for rundown of the machines is provided via the 9 bar bus that is connected automatically.
- Two separate panels to control the gas seals of each compressor train.

The anticipated advantages of using dry gas seals which were confirmed in the course of plant operation are:
- Buffer gas system no longer required,
- no penetration of oil into the process,
- reduced current expenditure for maintenance, service intervals are no longer determined by the sealing system,
- energy savings.

As a consequence of these positive experiences the second untreated gas train was revamped one year later and in addition the ethylene and propylene compressors (with tandem seals). A new recycled gas compressor ordered, has been completed with duplex dry gas seals.

After 7 years of using gas seals in that plant, alongside the major advantages, only small disadvantages were noted. Higher expenditure in the event of repairs, due to long delivery spare parts require storage and this causes storage costs, higher susceptibility of the seal to dirt and humidity, if the compressor is opened the seal must be replaced by a new seal

C603/029/2003

Radical upgrading of ethylene plant turbo-machinery control and protective systems

N H WELLS
Huntsman Petrochemicals (UK) Limited, UK

1. INTRODUCTION

This paper describes the experience of one end user in retrofitting modern electronic controls and protective systems to the main compressors of an ethylene production plant.

The paper describes the background to the project, development of the project scope, and explores some of the issues raised by implementing a radical change on a plant constrained by onerous reliability and availability requirements.

2. BACKGROUND TO PROJECT

The Huntsman Olefins 6 plant at Wilton in the north-east of England is one of the largest petrochemical production units in Europe, producing approximately 3,500 te/day of gaseous hydrocarbon intermediate products in a single stream plant.

The plant's cryogenic gas separation process is supplied by a single very large gas compression train, which is driven by multi-megawatt steam turbines. Similarly sized propylene and ethylene compressors provide refrigeration. The three compressor trains and four steam turbines are closely integrated into the plant. In addition to the compressor integration, the steam turbines are driven by by-product, high-pressure steam raised on the main process furnaces. Three of the turbines exhaust steam at intermediate or low pressure for use in the process.

Any interruption to machine operation causes both immediate loss of production, and the flaring-off of unprocessed gas at a rate of several hundred tonnes per hour. This is a major financial loss, and also has environmental implications due to smoke, noise and light generated by flaring. Hence machine reliability is of paramount importance: this is reflected in conservative selection of machines and associated systems, and in the installation of a high level of protective trip systems designed primarily to protect the machines from damage due to plant or control upsets.

In addition to basic design, reliable operation requires continuous attention to achieving the highest standards of operation, monitoring, and maintenance. A previous paper (1) described a 'machinery care' process applied to improving machine reliability on ICI Teesside sites. This process was initially developed on Olefins 6, then under ICI/BP ownership, and has continued to be developed and applied. Very high operational reliability has been achieved, with no significant machine breakdowns in ten years of operation, and machine on-line availability running typically at 99.9% of plant on-line time. This has been achieved while simultaneously extending overhaul intervals from 3 to 5 years. However, the small residual non-availability has a disproportionate impact on plant production and flaring. Following a review of operating history, it was concluded that hardware upgrading would be required to achieve further improvements.

3. PROJECT CONCEPT

As outlined above, the primary driver for change was the environmental and economic benefits resulting from reducing trip frequency and increasing on-line availability.

Detailed trip records were available, covering over 20 years of operation, providing root cause and production loss data for every machine outage. These records were analysed in detail. Figures 1 and 2 show the incidence and duration of machine trip events. A more detailed analysis was undertaken, classifying trip events on the basis of root cause. Four categories were identified: -

- Plant-induced trips (i.e. genuine demands)
- Trips due to human error
- Machine failures
- Machine support system failures

The results of this analysis are shown in figures 3 and 4. Several conclusions were drawn from this work: -

- The incidence of genuine trip demands is relatively low, representing approximately 25% of the total.
- Trips due to human error and testing of trip systems represent a further 22% of the total, reflecting efforts to improve reliability through systems and training
- The incidence of genuine machine failures is very low (less than 10% 0f total), although the duration of these outages is disproportionately long.
- The incidence of trips due to machine support system failures is very high (over 40% of total), and has increased with time.

Further analysis of machine support system failures showed a wide variety of root causes, including turbine governor and extraction control system faults, compressor anti-surge system faults, and spurious trips due to faulty trip system initiators.

4. EXISTING MACHINE CONTROLS/DEVELOPMENT OF PROJECT SCOPE

4.1. Speed/Extraction Controls

Turbine speed is controlled by conventional mechanical governors, gear-driven from the turbine rotor, opening turbine inlet valves (V_1 valves) via a hydraulic amplifier, pilot valve,

C603/029/2003 © With Author 2003

and power cylinder. The speed set point to the mechanical governor is transmitted pneumatically from a local control station, which in turn is driven from the main plant Distributed Control system (DCS) to control compressor suction pressure.

The two larger turbines are two-section extraction/condensing machines with a second set of valves (V_2 valves) controlling steam admission to the condensing section of the turbine. Simultaneous control of machine speed is achieved with a mechanical 'three arm linkage' interposed between the governor hydraulic amplifier and the pilot valve of both V_1 and V_2 valves in such a way as to vary extraction steam flow-rate while holding turbine power and speed constant.

Historical reliability of the governor itself has been relatively good, with one complete failure and a number of minor malfunctions in the last 20 years. However, the associated three-arm linkage and local pneumatic control stations have been a potent source of unreliability, particularly in recent years. The three-arm linkage is a particularly complex mechanical system of levers, sliders and links, is prone to wear in service, and difficult to set up correctly. There is a shortage of skilled personnel capable of fault diagnosis and set-up. As a purely mechanical device, there is little scope for integration with the plant DCS, leading to control problems when operating close to turbine constraints.

Electronic governors (and extraction controls) were originally introduced in the 1970's, and have been developed to be standard equipment on large steam turbines for process applications. They have the added attraction in this application of eliminating local pneumatic control stations, and replacing three-arm linkages with simple algorithms linking speed and extraction control functions with output to valves. Hence installation of electronic governors and extraction controls formed a central plank of this project from the outset. 'Redundant' control was also seen as essential to achieve the required reliability.

4.2.   Compressor Anti-Surge Controls

Compressor anti-surge protection is provided by recycle valves, operated by simple local pneumatic flow controllers, set to open the recycle valve to maintain a constant minimum flow-rate through the compressor. This configuration has been seen as robust and reliable, providing adequate protection on a plant typically operating at steady high throughput. However, analysis of trip history showed: -

−   A significant number of machine upsets due to anti-surge system faults, mainly failures associated with the local pneumatic controllers.
−   A significant number of incidents in which a plant upset was caused or exacerbated by the failure of an anti-surge system to provide adequate protection when required.

Modern electronic anti-surge controllers offer potential improvements in both areas of concern, through elimination of local pneumatic control stations, and enhance controller functionality, including speed compensation.

Generally, specialist machine controls vendors offer both turbine speed and compressor recycle controls, these two elements were combined into a core 'specialist machine controls' package.

4.3. Compressor/Turbine Trip System

The machine protective systems have a large number of initiating devices, inputting to relay based trip logic to shut the turbine down via solenoid operated oil dump valve and hydraulically-operated trip valve in the main steam supply to the turbine.

The original system was engineered with a heavy emphasis on 'failsafe' (rather than operational) reliability, and incorporates unusual features, including twin parallel steam trip valves, and provision for full-load on-line testing of trips right through to the final trip element. Trips are mainly single-channel switch type devices, configured to fail safe; initiator faults result in an immediate machine trip. Historically, trip system faults are the largest single source of spurious machine trips. Hence trip system upgrading formed an essential element of the reliability project.

Three essential improvements were seen as necessary: -

– Conversion of all trip functions to a fault tolerant arrangement, based on multiple initiators and voting logic.
– Conversion of most trip initiators to transmitters, rather than digital (switch) devices, to facilitate on-line monitoring of the 'health' of initiators.
– Replacement of relay logic with a programmable electronic logic (PLC) device capable of handling the increased number of inputs. Safety PLC's are available from a range of manufacturers, and are approved for use in safety-critical applications including boiler burner management and nuclear power.

In addition to the three main project elements above, a number of smaller items for improvement were identified: -

4.4. Over-speed trip System

The existing electronic two-out-of-three voting over-speed trip system had performed reliably, but was based on obsolete speed measuring and switching units. Consequently a replacement electronic over-speed system was included in the project scope, as part of the specialist machines controls package.

4.5. Oil System Controllers

Oil system pressure and temperature is controlled with local pneumatic controllers. Recent operating history has been good, however, the potential of these items to cause machine trips and upsets has been demonstrated in the past. Consequently, the project scope was extended to include elimination of local oil system controllers, by transferring control functions to the plant DCS.

4.6. Local Instruments/Operator Interface

Historically, machine-related display instruments, alarm/trip annunciation, and machine pneumatic control stations have been housed in control panels local to the machines. Since a significant proportion of these items were due to be replaced as part of the main machine controls upgrade, it was decided that all the remaining panel instruments should be transferred to the plant DCS, and also that the DCS should be adopted as the interface for all the machine control and trip packages. This allowed the local panels to be eliminated completely, being replaced with local DCS terminals, allowing all machine-related functions to be accessed from a single point local to the machines.

The outcome of this decision was a requirement for a significant expansion of the plant DCS, including an additional central processor and communications enhancements. Although this added significantly to the cost of the project, it was seen as essential to realising the full potential benefits of the upgrading, and considerably simplified the problems of operator training.

4.7. Turbine Start Sequences

The original mechanical speed governors operated over a limited speed range, typically 66-100% of maximum. During start-up, the machines are operated at reduced speed under manual throttle control to allow the turbine to warm up in a progressive and controlled manner.

Electronic governors have the capability to control speed from just above standstill, creating potential for automation of start-up. Since the complexity of machine controls coupled with the low frequency of machine start-ups have perennially raised issues of operator training, experience, and skill, automation of start sequences was seen as a highly desirable feature, and was adopted as a key requirement of the project.

Separate hot and cold start sequences were required.

5. PROJECT IMPLEMENTATION

5.1. Structure/Implementation programme

Within the scope developed in Section 4 above, the project sub-divided into four coherent packages: -

A 'specialist machines controls' package, embracing turbine speed/extraction controls, compressor anti-surge controls, and turbine over-speed trips.

A 'machines trip system' package.

A 'DCS expansion' package, including the necessary hardware, and programming effort to transfer controls and operator interface onto the plant DCS. This included programming of start sequences.

A general 'control/electrical package', covering all required instrument/electrical work outside the three vendor packages, including cabling, additional/replacement initiators and monitoring devices, installation of electronic valve positioners and converters, etc. This package was designed and constructed by the overall project managing organisation, which also specified and coordinated testing and supply of the three vendor packages.

5.2. Equipment Selection

5.2.1. Specialist Main Machine Controls Package

A (small) number of potential vendors were identified, with experience of supplying machine controls engineered for variable speed turbine driven process compressor applications. Vendor offerings differed considerably in equipment configuration, and technical emphasis, which ranged across compressor control applications, turbine control applications, and hardware capability. Some were more broadly based than others.

Hardware configuration ranged from discrete single/dual loop controllers, with automatic switching from main to back-up controller in the event of a fault developing, to larger PLC-based units incorporating multiple control functions within a single triple-redundant controller. In one case, integration extended to combining all control functions for four machines into a single unit.

Hardware reliability data was requested, and supplied by all vendors. Although there were relatively large differences in quoted reliability, impact on overall reliability was concluded to be small. Overall reliability is dominated by measuring and control elements common to all control configurations.

The discrete controller option was ultimately selected, based primarily on previous application list, also on the perceived strength and accessibility of the vendor's field support organisation.

5.2.2. Machines Trip Package
As with the specialist controls package, a number of potential vendors of high-integrity PLC-based trip systems were identified. The range of technical options was considerably narrower, however. All systems were based on either triple-redundant or dual-channel, twin processor per channel arrangements, with automatic fault diagnosis and switching between channels. With vendors offering essentially the same performance, selection was based on price, and compatibility with the existing plant DCS system.

5.3. Development of Start Sequences
The concept of automated turbine start sequences was identified as a critical success factor for the project at an early stage. It was recognised that automation had the potential to radically simplify turbine start-ups. It also rapidly became clear that automated sequences were the key to 'seamless' integration of machine controllers with the plant DCS system. However, sequence development proved to be one of the most demanding elements of the project, both conceptually, and in implementation to a timetable.

To launch the process of developing start sequences, the current stages of manual start-up for each machine were identified, and tabulated, with critical machine parameters for each stage, with limits of acceptance for each parameter, together with action to be taken in the event of non-conformance. Discussions with potential controller vendors soon demonstrated that none of the vendors had suitable software available 'off the shelf'. The selected speed/extraction controllers were unable to offer the required functionality within the controller, hence an early decision was taken to implement the start sequences within the DCS, as part of the DCS expansion package.

Development of the sequence code proved a major challenge for the specifying organisation, the supplier, and the end user, due to the combined complexity of the required functionality, limitations/features of the turbine/compressor controllers, and the limitations of existing turbine hardware, which was largely manual in operation. Time and resource requirements were significantly under-estimated, which impacted on the 'integrated FAT' test schedule, ensured that significant programme development went on throughout the equipment testing programme, and that final acceptance testing was carried out at a very late stage prior to the final 'during overhaul' phase of project implementation and commissioning.

C603/029/2003 © With Author 2003

Although (not surprisingly) some 'bugs' in the automated start-up sequences continued to emerge right up to and during machine start-up, the sequences worked well during commissioning, and drew positive comment from operating personnel.

6. ENGINEERING FOR RELIABILITY

Since the primary objective of the project was to improve reliability, much effort was directed towards this end at all stages of the project. A number of specific measures were adopted:-

6.1. Equipment Configuration

As far as possible, redundant fault-tolerant configurations were adopted for all equipment. For new equipment, the configurations adopted are described in section 1 of this paper. In addition, the operating configuration of the existing turbine hardware was modified to permit continuous operation with twin parallel steam trip valves, eliminating the risk of spurious trips due to mechanical malfunction of the trip valve, a significant source of past outages.

6.2. 'Nested' Programme of Vendor Testing

A structured and progressive programme of testing vendor packages was adopted. Each package was subjected to Factory Acceptance Test (FAT) before leaving the vendor's works. Equipment was functionally tested to the fullest extent possible at this stage, including simulation of faults, power supply failures, etc.

Subsequently, an 'Integrated FAT' was carried out in the factory of the DCS vendor, combining the machine control, trip system, and DCS expansion packages. This proved extremely valuable, allowing extensive testing of communications, and also development/testing of turbine start sequences, including simulated turbine starts, etc.

The 'Integrated FAT' was repeated on site, following installation in the plant control building.

Finally, following full installation, cabling, and loop checking of individual field connections, a full programme of trip testing and control loop functional checking was adopted.

6.3. Design Review Processes

Proposed designs were subjected to a number of formal review processes, including Hazard and Operability and detailed Failure Mode and Effect studies. The machine trip system was engineered using IEC 61508 methodology.

In addition, all instrument/electrical design was subjected to very detailed informal scrutiny, and many changes were incorporated at this stage. Many of the improvements were directed at improving facilities for on-line fault-finding and equipment change-out.

Turbine control mechanical modifications were similarly scrutinised, and a number of improvements proposed and incorporated.

6.4. Field Testing of Instrument Loops

All field instrument loops were checked in stages. Following 'cold' continuity checks, equipment was made live, and a prescribed programme of 'hot' checks were carried out by the installation contractor prior to handover to operating personnel. Following handover, a full programme of functional testing of trip and control systems was carried out by a combined team drawn from installation contractors and operating personnel. Where this uncovered a need for modifications or remedial work, a final 'clean' test was carried out.

6.5. Engineer and Technician Training

Training was recognised to be a critical determinant of project success. Key engineering personnel were trained by the main vendors prior to equipment arriving on site.

In-house training packages were developed and delivered to operating and maintenance technicians. Following site acceptance tests, controls, trips and DCS interface were operated in a stand-alone, 'simulation' mode which as used to provide hands on training for operating technicians, including realistic simulated machine start-ups using the start sequences.

In addition to the general pre-overhaul training of all operating personnel, a number of team leaders worked in the machines area during the plant shutdown phase, and were intimately involved in project commissioning.

7. PROJECT TIMETABLE

Project concept was first discussed in the spring of 2000, and a preliminary scope was put together that summer.

Detailed engineering started in September 2000, and a detailed engineering proposal was competed by March 2001. The project was sanctioned in May 2001, leaving twelve months for detailed design and implementation. Vendor packages were ordered for delivery in the period Oct/Nov 2001, with a view to a site installation and testing period of approximately six months prior to plant shutdown and final installation and commissioning of field instrumentation.

The introduction of an 'Integrated Factory Acceptance Test' and delays associated with development of start sequences resulted in all equipment coming to site in late January, however intense construction activity and programme re-scheduling allowed the Site Acceptance Test to proceed to plan in late February.

Following completion of the Site Acceptance Test, the equipment was used for operator training until just before final installation and commissioning.

Final installation and field commissioning occurred in May/June 2002, during a full maintenance shutdown of the plant. During this intense six week phase, all field instruments were installed, connected and tested, in parallel with overhaul and modification of the main machines, followed by progressive commissioning as part of the overall machine and plant start-up programme.

8. COMMISSIONING AND STEADY OPERATION

Historically, the overall plant commissioning programme is built around machine commissioning, with an initial mechanical over-speed test of each turbine uncoupled from its compressor, followed by start-up of refrigeration machines, and ultimately start-up of the process gas machines.

Although the need for uncoupled test runs should ultimately be eliminated following installation of reliable electronic over-speed trip system, and removal of mechanical over-speed trip, they were retained as part of the project programme, both as a 'one off' proof of the effectiveness of the new over-speed trip system, and to allow the project to be commissioned in stages.

The over-speed runs also provided a first opportunity for 'hands-on' training of plant personnel in machine start-up using start sequences and new controls.

Progressive commissioning proved very valuable when problems were encountered with speed measurement at the outset of over-speed trip testing, both electrically (voltage drop across barriers) and mechanically ('shrouding' of probes within steel housing), however both problems were corrected with no delay to the overall commissioning programme.

Similarly, start sequences were tested, and a number of faults corrected prior to coupled start-up.

Final coupled start-up was generally straightforward, with operating personnel adapting rapidly to the new control interfaces, and finding the start sequences much easier to follow than the previous manual arrangements. Further start sequence faults emerged and were corrected during the start-up.

Initial plant operation was unsteady, and some re-tuning and adjustment of anti-surge controllers was required, eliminating a tendency to open fully when faced with a relatively minor perturbation.

Subsequent plant operation (three months) has been very steady, with one project-induced short-term machine outage due to as faulty electrical connection.

9. CONCLUSIONS

9.1. Full re-instrumentation of an ethylene plant main machine installation is a major undertaking, however given time and adequate resources, it can be achieved without loss of availability or reliability.

9.2. Success is critically determined by attention to detail at every stage of design, installation, and commissioning, supported by rigorous checking, and thorough training of personnel.

9.3. The level of resources required for reviewing and checking was consistently greater than had been envisaged initially. This was true at every stage of the project, but most obviously so during final loop checking and live trip testing.

9.4. Similarly, the resources required for training considerably exceeded initial estimates, and despite major effort in this area, post-commissioning training has proved to be necessary to further boost operator confidence and competence.

9.5. Despite rigorous efforts at every stage of the project, structured testing, and a high level of commitment and teamwork from personnel throughout, faults emerged at a late stage, and one post-commissioning fault has emerged resulting in a brief production interruption. Given the complexity of the project, this is almost inevitable, and should be recognised when projecting benefits of future projects.

10.   REFERENCES

10.1.   Wells N H; Machinery Care- an operator's approach to reliability.   IMechE Paper C508/015/96.

Figure 1:-   Incidence of machine trips, 1982 - 2002

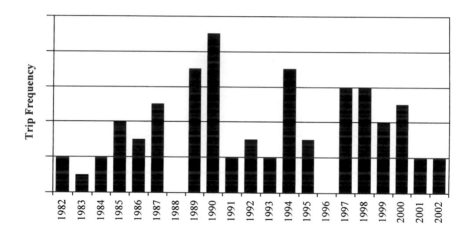

Figure 2:-
Duration of machine trips, 1982 - 2002

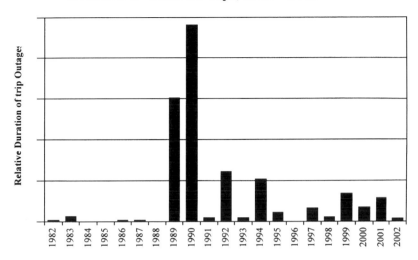

Fig 3 Trip Incidence by Root Cause

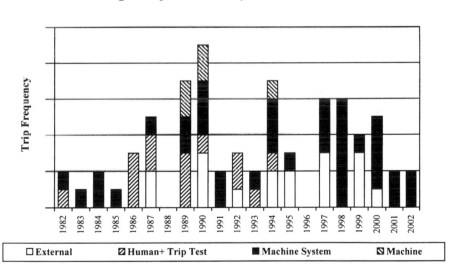

Fig 4 Duration of Trips by Root Cause

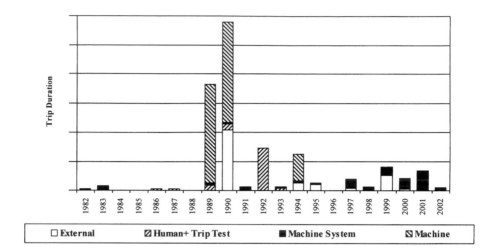

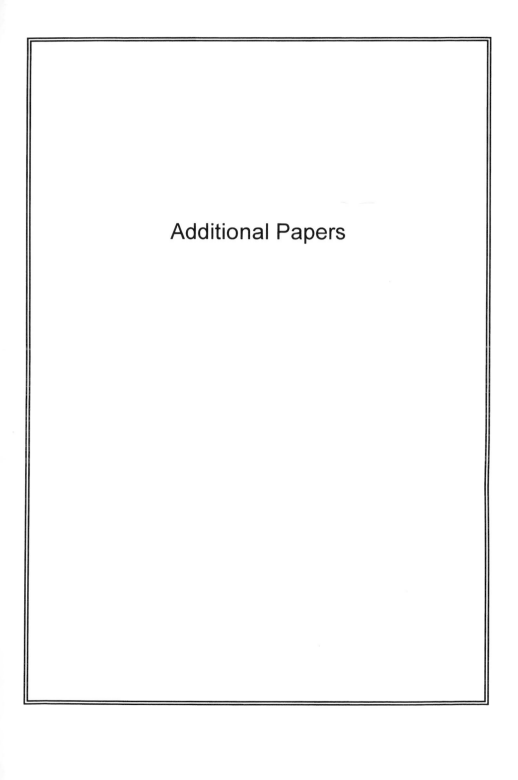

Additional Papers

C603/030/2003

Design consideration for helico-axial multiphase pumps

M CORDNER and J DE SALIS
Sulzer Pumps, UK

ABSTRACT

Multiphase pumps have now been deployed to facilitate oilfield development and production enhancement at several installations. At this conference in 1999 the same authors presented a paper on the status of helico-axial multiphase pumps at that time. This paper will:-

− Present a brief update on the installations described in the previous paper.

− Review key multiphase pump design parameters with respect to both the pumping equipment and operational influences,

− Highlight how the pump design has been influenced by experience derived from installations in demanding environments with particular reference to the multiphase pumps installed on the Dunbar Platform in the UK sector of the North Sea.

1.0 UPDATE ON 1999 PAPER

In the 1999 paper "Multiphase Pumps: Helico-axial Technology Applications and Developments"[1], the helico-axial technology as derived from the Poseidon project was introduced and two applications in particular were described in detail. These were two onshore pumps in the Samatlor Oilfield Western Siberia and two pumps installed on Total's Dunbar platform in the North Sea.

1.1 Samatlor Field

The two multiphase pumps (each rated 83,000 bpd) installed on the Samatlor field were commissioned in the spring of 1998 since when they have both been running continuously. The pumps, which run in parallel, have operated successfully and reliably with no major problems although on one occasion a fault in the lube system led to a thrust bearing failure (which was repaired within a few days).

Further to the successful introduction of these multiphase pumps, two additional pumps have since been installed to implement a further development of the same field (Figure 1). These new pumps are piped to the same processing facilities but are located further away and consequently have a higher differential pressure and driver rating. See Table 1 for details. The two new pumps were commissioned in the autumn of 2001, thereby again enabling the oil and gas mixture to be boosted to the existing treatment facilities and avoiding the need for a new infield treatment plant.

Table 1

		Samatlor 1	Samatlor 2
Distance from wells to pump		5 to 25 km	1 to 10 km
Distance from pump to process facilities		15 km	25 km
Pump total capacity	(bpd)	83,000	150,000
Inlet pressure	bar abs	5 to 10	5 to 10
Discharge pressure	bar abs	≤ 18	≤ 30
Inlet GVF (Gas Void Fraction)	(%)	40 to 88	40 to 90
Pump frame size		MPP 7	MPP 7
Pump speed	(rpm)	1500 - 4000	3000 - 5750
Driver rating	(kW)	400	2000
Operating hours		63,000	12,000

The scope of supply is similar except that in the case of Samatlor 2, which has a significantly higher rated power, the variable speed facility is achieved by means of a variable frequency inverter rather than a fluid coupling. The power absorbed by a multiphase pump at reduced speeds does not necessarily follow the conventional square law curve (Figure 2) because of the transient nature of the operating conditions. Consequently at reduced speeds the power absorbed may be higher than would be indicated by this law. The VFD was therefore selected to optimise the overall efficiency of the complete train.

1.2   Dunbar
Two large vertical multiphase pumps, each rated for a total capacity of 1,200 m^3/h (180,000 bpd) and driven by a 4.5 MW motor entered into service on the Dunbar platform at the end of 1999. As regards suction, discharge and differential pressures, these pumps are at the leading edge of multiphase technology. The manner in which the pumps operate at different pressures and the way they have been adapted to suit changing field conditions is believed to be unique. A detailed analysis of the experience gained by the operator was presented by Total Fina Elf at the SPE conference in 2001[2]. This paper concludes that after some commissioning issues were resolved, mainly with the mechanical seals and systems (lube oil pumps and coolers) the pump installation has been a success. The paper also explains how the hydraulic flexibility of the pump installation has enabled the operator to manage the changing characteristics of the field, a theme which is considered further in this paper.

1.3   Other projects
Since 1999 other helico-axial multiphase pump installations include:-

–   The Priobskoye field in Russia (owned by Yukos Oil Co) where the largest multiphase pumps in the world (driver rating 6600 kW) were commissioned in the summer of 2001

–   Hassi-Berkine in Algeria for which two 2500 kW pumps have been supplied to Anadarko

- Abqaiq field in Saudi Arabia where a multiphase pump was commissioned in June 2001

2.0 DESIGN CONSIDERATIONS

The fundamental requirement for a multiphase pump is that it must be capable of acting both as a pump and/or a compressor, which makes significant demands on the pump design in many respects. This paper will review four aspects of multiphase pump design and the manner in which these have been influenced by increasing experience of pumps operating under different conditions.

- Mechanical Design

- Hydraulic Design

- Influence of system and station layout.

- The ability of the pump to adapt to fluctuations in operating conditions, as regards both short term transients and long term field evolution

2.1 Mechanical design considerations

2.1.1 Pump design concept
As regards the mechanical design, the pump is of the multistage barrel casing design (Figure 3) with a withdrawable cartridge comprising an axially split inner casing incorporating conventional mechanical seals and bearings. The design is generally in accordance with API 610, although the rotordynamic design is based on the API 617 specification for compressors. A pressurised seal system (similar to a plan 53 system) prevents gas leakage or solid particles contaminating the bearing area and ensures that a liquid film is always present at the seal faces even under dry running conditions. The lube oil system for the bearings can be either separate or integral with the seal system.

In practice the main areas for concern, and most likely to give rise to problems, are the same as for conventional pumps i.e. mechanical seals and bearings. Other aspects of the design peculiar to multiphase pumps such as the rotordynamic design and sand handling capability where problems might have been expected to occur have proved not to be problematic.

2.1.2 Mechanical seals
In the case of conventional pumps, a mechanical seal is usually only required to operate at a single point or a small number of clearly defined duty points, usually under relatively stable conditions. However with a multiphase pump, the seals are required to operate under a wide range of constantly fluctuating pressure/temperature combinations, which has certain implications for the seal design.

For example in the case of the Abqaiq pump in Saudi Arabia the suction pressure has been observed to fluctuate between 3 and 17 bar.a, perhaps 2 or 3 times per minute under slug flow conditions. Seal leakage was greater than anticipated. However this did not impair the operation of the pump, no seals failed and investigation revealed no evidence of undue wear at the seal faces. Modifications to the mechanical design and metallurgy of the seal were nevertheless introduced and these are currently being assessed.

As regards mechanical seals, the most arduous installation of which Sulzer has experience is the Dunbar Platform. Operating conditions are as follows:-

Pump Capacity	up to 180,000 bpd
Suction Pressure	50 to 125 bar g
Discharge Pressure	up to 125 bar a
Pump Speed	3500-6000 rpm
GVF	30-90%
Inlet Temperature	20-100°C

The reduction in suction pressure and increase in GVF are due mainly to the ongoing reduction in pressure in the wells.

As noted by Total Final Elf[2], blistering occurred on the mechanical seal faces after about 800 hours operation. This was caused by higher than anticipated temperatures at the seal faces. The carbon seal faces were therefore replaced by resin impregnated carbon seals after which no further problems occurred. The degree of balancing at the seal faces was also modified.

Problems were also encountered with the pumps on the seal oil system due to the ingress of air and the fact that the pumps proved incapable of handling the low lube oil viscosity. Pumps from a new supplier have been installed which has improved the situation.

2.1.3 Bearings

The selection of bearings is based on conventional criteria, but with particular emphasis on the need for journal bearings with high stiffness to achieve the required rotordynamic characteristics. Consequently anti-friction bearings are fitted on low speed pumps and tilting pad journal bearings on higher speed pumps.

The axial bearing for both designs is of the tilting pad type. Axial thrust is unidirectional although provision is included for reverse thrust.

For high differential pressures Sulzer have developed a balance piston, which means that significantly increased differential heads can now be achieved, and the pump operating envelope extended. In the case of the Dunbar pumps for example, differential pressures of up to 80-85 Bar have been achieved. The balance piston reduces the axial thrust and also reduces the pressure at the inboard mechanical seal enabling the same seals to be fitted at both ends of the pump, both at suction pressure. The balance piston is specially designed for multiphase pumping, being profiled so as to maximise rotor damping and minimise leakage.

2.1.4 Rotordynamics

A key challenge in the development of multiphase pump was the rotordynamic design in the context of operation between 100% liquid and 100% gas with reduced stiffness and damping at the interstage regions of the pump. The rotordynamic design of the pump is derived from experience with both pumps and compressors and is assessed with respect to the API 617 standard for compressors. High speed pumps run between first and second critical speeds which means that the bearing design must have sufficient stiffness to enable the pumps to safely pass through the first critical speed when running up.

To date this approach has been validated by the successful rotordynamic performance of multiphase pumps in field conditions.

2.1.5 Sand
The presence of entrained sand is frequently a characteristic of multiphase applications. There are two issues associated with sand, i.e. erosion within the pump and the integrity of the mechanical seals. Sulzer have carried out a testing programme in conjunction with the Institut Français du Pétrole (IFP) at their test facility at Solaize to investigate the impact of sand content in the oil/gas effluent being pumped. This has made it possible to determine the nature and location of wear in the hydraulic passages and to develop an appropriate procedure for coating the relevant areas. The pump also incorporates design features to prevent the ingress of sand into the mechanical seal area.

Furthermore the helico-axial pump design is inherently very tolerant of sand because of the open hydraulic passages and relatively large clearances. This has enabled pumps to operate when protected only by typically a 3 or 4 mm coarse strainer and pumping effluents containing up to 300 ppm of sand.

2.2   Hydraulic design

2.2.1   Pump design
The Sulzer range of multiphase pumps is derived from the Poseidon helico-axial concept (second generation hydraulics developed by the Institut Français du Pétrole).

The pump is of the multistage design, where each compression stage (Figure 4) consists of an impeller, mounted on a single rotating shaft, followed by a fixed diffuser. The impeller blades have a typical helical shape. The profile of the open type impeller and diffuser blading arrangement is specifically designed to prevent the separation of the gas liquid mixture during the compression process. Within the pump the hydraulic design of these compression stages is adapted to take account of the decreasing volumetric flowrate due to the compression of the gas content. The continuously open hydraulic passages accommodate sand and solid particles in suspension and special care has been taken in the design of the pump to prevent their accumulation in the pump casing.

2.2.2   Hydraulic performance
As stated above, the pump hydraulic design, hydraulic performance and corresponding curves have all been derived from the design which emerged from the Poseidon Project. Feedback from various site installations has in general demonstrated a good correlation between predicted curves and curves based on site measurements. For example Saudi Aramco have stated that in the case of the pump installed at Abqaiq "pump performance was on target or better....compared to manufacturer predicted values". The efficiency was also better than predicted and Aramco have observed that "In general, it can be stated that the pump performance curves adequately represent true pump performance" [3].

2.3 Influence of system and station layout

Experience with various installations has demonstrated the importance of the initial multiphase analysis or modelling of the system and the design of the station layout.

The rated duty points cannot be considered in isolation. The nature of the application dictates that key operating conditions such as the suction pressure and gas content are always transient, with inherent liquid and gas slugs being characteristic features. These are determined by phenomena external to the pump i.e the well conditions, the process pipework and the station layout.

2.3.1 System design

In the first instance there is inevitably a degree of uncertainty regarding the flow rates and pressure predictions. The design of production facilities is generally based on extrapolations from the results of exploration well tests, delineation wells and reservoir characterisation. Consequently when the production wells are drilled and completed, the actual production data may differ from the prediction. Sulzer have experience of one case where the GVF was predicted to be about 60%, but was actually found to be nearer to 90% when the pumps were installed on site. This new situation was however still within the design capability of the pump as originally supplied.

Similarly inaccuracies or uncertainties in the flow modelling program (input data / modelling program selected) may mean that system losses are not as predicted.

Well instabilities especially under "end of life" conditions can also contribute to changing flow and pressure conditions at the pump suction[4]. Such instabilities are characterised by strong fluctuations in GVF with a periodicity, which can vary from 15 minutes to several hours.

Another factor is the number and type of wells connected to the pump. When there are several wells connected to the pump (e.g. about 20 as in the case of Dunbar) then pressure variations in individual wells tend to be dissipated within the overall flow whereas a small number of wells is less able to benefit from this "averaging" effect and so is more likely to give rise to a slug flow regime.

2.3.2 Layout

It is important not to underestimate the importance of the layout and the consequent impact upon the pump and system. For example a large diameter suction pipe is usually considered advantageous because of lower suction velocities and losses. However this may have an adverse effect in that the greater velocity differential between the gas and the liquid can result in slug generation.

On the discharge side the pipe diameter is a key factor in determining the pump discharge and differential pressure and consequently the pump selection, the number of pumps, power requirements and corresponding costs. The discharge pipework can also influence the pump operation. In one case[3] an unstable process and severe slugging at the pump were alleviated by switching the pump discharge from a 10" production line to a 16" production trunkline. Bends with insufficient radius and vertical rising sections in the discharge pipework can also lead to unstable flow thereby inducing vibration and slug flow within the station.

C603/030/2003 © With Author 2003

Consequently the design and modelling of the system should preferably be carried out in conjunction with the pump manufacturer so that the impact of the layout and pipe diameter etc on the pump can be assessed. The design of the pump package can then be reviewed and adapted in the context of this information especially as regards slug flow conditions.

Other elements of the layout include a station bypass line (when provided) which enables production to proceed (albeit at a reduced rate) when the pump is not available. Also in certain circumstances (for example to provide minimum flow protection for a pump which discharges into a high pressure discharge line) a recycle line will be incorporated into the station facilities.

2.4   Field condition evolution

In addition to uncertainties arising from the system design, it is often necessary to be able to adapt the pump to take account of changes arising from field evolution or due to different operational strategies.

– The production profile of the reservoir and consequently the demands on the pumps will change over the life of the field, as in the case of the Dunbar field which is considered in more detail below.

– The number of wells and the characteristics of the wells connected to the pump may vary

2.5   Tools available

A number of tools are available to enable the pump manufacturer and operator to respond to the demands arising from these transient conditions and process design uncertainties.

For short term transient conditions.

– In view of the short term process uncertainties and long term field evolution it is advisable to provide a generous margin between the pump absorbed power and the driver rating, typically 20-25%.

– Variable speed operation provides a high degree of flexibility, either by means of a fluid coupling or a frequency inverter.

– A flow distributor may be installed immediately upstream of the pump. This serves various functions; in particular it serves to distribute the flow evenly between pumps operating in parallel. As a static mixer, it smoothens out fluctuations in the GVF at the pump inlet and dissipates the energy from liquid slugs. Furthermore the flow distributor is part of the (optional) integral flow measurement system (described below). At the present time we have experience of a single flow distributor serving up to four pumps working in parallel, (Figure 5) although if the pumps are operating at different suction pressures (such as on the Dunbar platform), then a dedicated distributor per pump is preferred.

For long term process changes / field evolution:

– The helico-axial pump is a multistage pump, incorporating typically up to 16 stages. The pump is of a modular design with a range of framesizes, for each of which 8 different hydraulic designs (analogous to specific speeds) are available. This means that the pump internal cartridge incorporating the hydraulic components can be replaced at the appropriate point in the life of the field to suit different flow or head requirements (similar to the retrofitting of a conventional barrel casing water injection pump).

– Further flexibility is afforded by either destaging a pump and/or varying the pump speed and/or using different numbers of pumps.

2.5.1 Control systems

The pump must be protected against high temperature and vibration in the same manner as a conventional pump. In addition various control strategies are available to the operator enabling him to control the pump to suit the production strategy. The pump itself can be controlled in several ways; including constant torque or power, constant speed, constant suction pressure or with respect to the discharge flow rate. The most common operating mode is constant speed with the speed being set by the operator at start up depending on the required capacity.

An optional feature of the control system is an integral flow measurement system. (Figure 6) This is achieved using the flow distributor, and a venturi on the pump outlet. During the pump factory testing, the level in the flow distributor is calibrated so that this level measurement can be used to estimate the Gas Liquid Ratio (GLR), which when taken together with information from the venturi, enables an approximate flowrate to be calculated. The software required for this calculation is incorporated in the control panel. Feedback from site installations has demonstrated the accuracy of this system, especially on the Abqaiq field where a good correlation was found between this system and measurements taken from an Accuflow multiphase flowmeter installed on the same site.

This flow measurement system (or an independent flow meter) can be used to facilitate discharge flow control which may be more appropriate for applications where process requirements focus on the flow rate or where surge control is of particular importance.

Again however the pumps cannot be considered in isolation. Consider for e xample a c ase where multiphase pumps are fed by downhole pumps installed in the wells. This means that any action e.g. to increase production by increasing the speed of the main pumps, is dependent upon the time required and the capability of the wells to respond accordingly.

The ability to measure the flow and adapt the pump speed accordingly means that the pump can operate in an unmanned station and be controlled remotely. This facilitates the installation of multiphase pump stations in areas where access is inconvenient or difficult.

3.0   CASE STUDIES

Two examples showing pumps handling a range of operating conditions are detailed below.

3.1   Abqaiq

In June 2001 a multiphase pump was commissioned for Saudi Aramco at the "Northnose" remote manifold in the Abqaiq field in eastern Saudi Arabia. This was essentially a trial test, the aim of which was to increase production in general and to attempt to use the pump to revive dead wells. This was achieved by adapting the configuration of the wells and adjusting the pump speed[3]. Initially two free flowing wells were connected to the pump suction. Then firstly one (Figure 7) and later both of the live wells were removed and gradually replaced by dead wells. The pump speed was also increased accordingly with the result that up to 8 dead wells are now back in production via the multiphase pump. The other main objective of achieving incremental oil production was also realised.

3.2 Dunbar Platform: Evolving Multiphase Pump Requirements
This example incorporates many of the above hydraulic design features and shows how the pump installation was designed to respond to evolving field conditions which have seen the suction pressure decline from about 120 bar.g to 50 bar.g and the Gas Liquid Ratio increase from about 3 to approximately 10.

The Dunbar platform in the North Sea is a satellite of Alwyn about 22 km away. (Figure 8) Oil and gas production is exported to Dunbar along a 16" multiphase pipeline. The pressure at the processing facilities on Alwyn is 70 bar which means that production must enter the export line at 125 bar. This was initially possible with natural energy but is now increasingly dependent on the multiphase pumps as pressure in the wells declines. There are about 20 producing wells at various pressures.

In order to match the pumps to the uncertainties of this evolving scenario, a segregation scheme was established. At present this segregated scheme comprises 3 flow lines i.e. a high pressure line for wells which are able to flow freely and two lower pressure lines, in each of which a pump is installed.

The production from high pressure wells, which does not need to be boosted passes directly through the high pressure line. Production from low pressure wells passes along the LP line which includes a booster pump designed to increase the pressure from 70 bar to 125 bar. Lower pressure wells are routed to the LLP line, in which the booster pump is designed to increase the pressure from 50 bar to 125 bar. There is also provision for a 3rd pump in serial/parallel configuration to maintain production from wells producing down to 30 bar.

This segregation scheme allows the natural energy of the wells to be harnessed for as long as possible, thereby minimising pump electrical power requirements (in contrast to a separator which requires the high pressure wells to be throttled so that the flow inlet from all wells to the separator is at a common pressure).

The key factor is the inherent flexibility of the pump design. The two pumps were each designed for an initial operating case as follows.

	LLP Line	LP Line
Pump type	MPP 8	MPP 8
No of stages	12	8
Total capacity	1071 m3/h (162,000 bpd)	780 m3/h (118,000 bpd)
Suction pressure	50 bar g	70 bar g
Discharge pressure	125 bar g	125 bar g
Nominal speed	5000-5200 rpm	4500-4700 rpm

Both pumps are capable of a maximum total capacity of 180,000 bpd and a maximum speed of 6,000 rpm. Both are driven by an electric motor rated 4,500 kW. Phasing in of the pumps has taken place as follows[2]:

- When the pumps were initially commissioned in 1999, only 7 wells were required to be boosted which was achieved using a single pump at low speed and a suction pressure of 90 bar g. (Figure 9)

- After 6 months this single pump was operated at a higher speed to suit a suction pressure of 70 bar.

- In August 2000, two pumps began to operate in parallel at two different suction pressures (i.e. 50 bar and 70 bar as per the initial design criteria).

- In the summer of 2001 the 8 stage pump was uprated to 12 stages (Figure 10).

- In the future, a third unit may be installed in series with one of the existing pumps to enable the wells to produce at pressures down to 20 or 30 bar.

An additional benefit derived from these pumps was that a temporary drawdown on the reservoir accelerated the recovery of the nominal flowrate of the wells after a long shutdown work-over.

The contribution of the multiphase pump is apparent from Total's prediction[2] that the amount of oil dependent on boosted production is estimated to increase from approximately 20% of the total production from Dunbar in 2000 to 84% in 2004.

4.0   CONCLUSION

Multiphase Pumping is now a mature technology, helping to prolong field life, increase production and generate revenue. As can been seen from this paper, the successful design of a multiphase pump installation involves attention to several design criteria; hydraulic, mechanical and process. Design approaches to these criteria have been validated in many areas by site experience, which has also confirmed additional benefits from multiphase pumping such as the ability to restart dead wells and to generate increased incremental production. Valuable experience and feedback from field installations is continuously received and evaluated thereby contributing to the ongoing development of the helico-axial pump design.

References

1. J. de Salis, M. Cordner : Multiphase Pumps; Helico Axial Technology Applications and Developments. I.Mech.E Seventh European Congress on Fluid Machinery, The Hague 15-16 April 1999
2. E. Leporcher, A. Delaytermoz, J-F. Renault, A. Gerbier, O. Burger : Deployment of Multiphase Pumps on a North Sea Field. SPE 71536 Annual Conference and Exhibition, New Orleans 30 September – 3 October 1996
3. S.G. Barlow, A.A. Hamoud, S.M. Al-Ghamdi : Field Testing of first multiphase pump in Saudi Arabia
4. J. de Salis, Ch. de Marolles, J. Falcimaigne, P.Durando : Multiphase Pumping – Operation and Control. SPE 36591 Annual Conference and Exhibition, Denver 6-9 October 1996

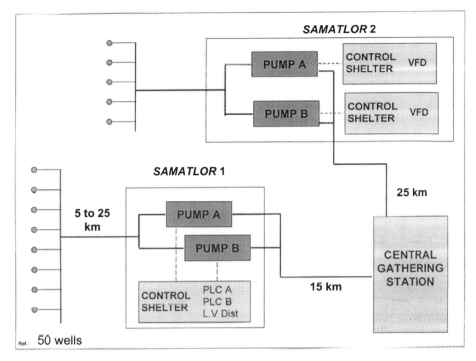

Fig 1 : Schematic Diagram : Multiphase pumps installed on the Samatlor field, Western Siberia

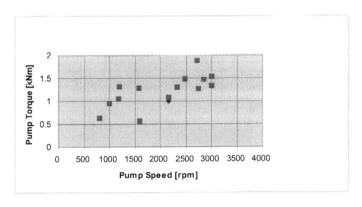

Fig 2 : Typical speed - torque curve

Fig 3 Helico-axial multiphase pump

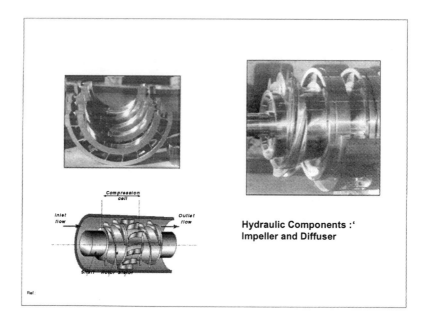

Hydraulic Components :'
Impeller and Diffuser

Fig 4 : Helico - Axial pump design

Fig 5 : Flow Distributor (total capacity 2m bpd) feeding 4 pumps in parallel

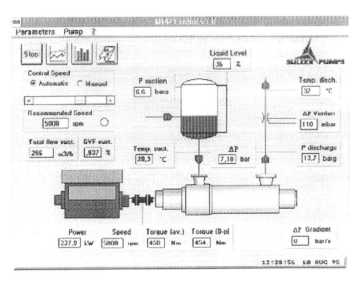

Fig 6 : Control Screen including flow measurement system

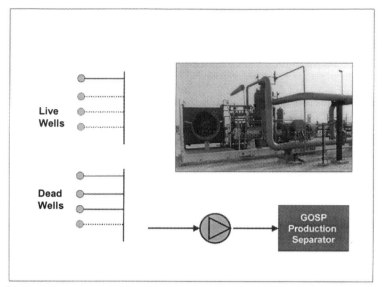

Fig 7 : Multiphase pump for Saudi Aramco Abqaiq field. Restarting dead wells

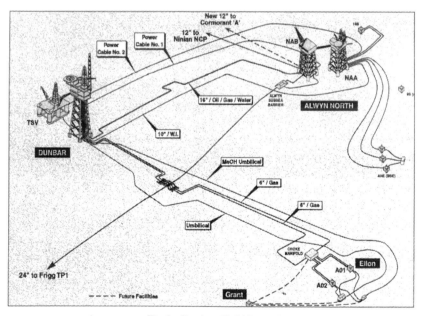

Fig 8 : Dunbar Field Layout

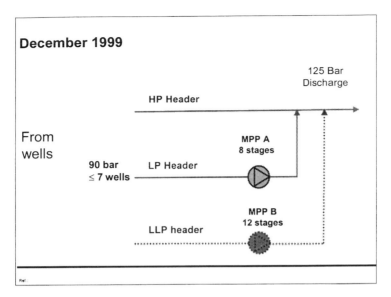

Fig : 9 Dunbar Flow segregation scheme

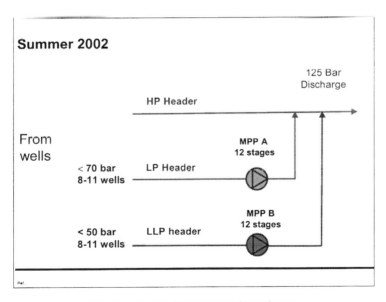

Fig 10 : Dunbar flow segregation scheme

C603/038/2003

Enhanced mechanical face seal performance using modified face surface topography

L YOUNG, E ROOSCH, and **R HILL**
Flowserve Flow solutions Division, Seal Group, UK

SYNOPSIS
Mechanical Seals have contradictory aims, to prevent fluid passage at a rotating machine shaft entry or exit and use the fluid to be sealed for lubrication, whilst maintaining sufficient life.
High performance mechanical sealing is provided by the use of laser-machined faces having both designed in waves and face tilt. This combination provides good sealing performance with a self-generated lubrication regime.
Seals using this technology and some duties are described, including a novel approach using surface tension of the fluid.

INTRODUCTION

The title of this paper could appear a little tautologous so it can be reduced to the potentially more manageable WAVY FACE TECHNOLOGY.

Over the last 5 decades mechanical seal design has involved the increasing understanding of the various physical effects and mechanisms affecting the condition of the seal faces and the consequent effect upon the fluid film between them.

Seal designers have to consider the pressure of the fluid to be sealed (normally arranged to be on the outer side of a mechanical seal) and the face deformations thus caused, as generally shown in the upper part of Figure 1.
Thermal deformation generated by sealed product temperature, flushing and face generated heat is shown by the lower part of Figure 1.
In addition consideration must be given to mechanical deformation created by drive loading, face fixation and torque resisting devices, adding complexity to the seal face and fluid film condition.

Needless to say the face materials, their strength, elastic and shear modulii, thermal properties plus their surface topography all add to this many faceted design conundrum.

The designer's task has been aided by the use of modelling methods, Finite Element Analysis and has been correlated with data obtained from ever more sophisticated testing work utilising advanced instrumentation and measuring techniques.

Previous work has used analysis based upon the calculation of pressure and thermal deformations; heat generation and using empirically derived functionality terms. A display of a seal face analysis prepared by such methods is shown in Figure 2. (1), (2).

The use of more advanced F.E.A. techniques and the latest laser micro machining developments have enabled seal designers to provide seal faces which generate lift with very little self generated heat, reducing the reliance on the lubricating properties of the fluid to be sealed and residual tribological effects of the face materials employed.

Circumferential waviness as a lift support mechanism by itself was explored more than thirty years ago, (figure 3). (3).

Adding radial tilt and a seal dam as a means of enhancing seal performance has been described in literature in a variety of applications (4), (5), (6), (7) & (8).

FACE DESIGN

Figure 4 shows the geometry of some wavy faces (highly magnified).
The shape has the form: -

$$h = \phi[Rd - R] \, Cos \, [n\theta]$$

Which specifies a wavy shape that is periodic where h is the depth down from the surface level at the dam (waviness amplitude), ϕ is the tilt angle, Rd is the radius to the seal dam and n is the number of waves.

Values of h could range from 2.5 to 10 microns, dependant upon the operating conditions.
The radial taper aspect of the face design provides hydrostatic support; face waviness promotes hydrodynamic load support during shaft rotation.
These actions promote non-contact operation of the seal faces, minimizing wear and heat generation.
The seal dam enhances leakage control by providing a restriction to the flow of fluid crossing the face.

Previous styles of face modifications may have led to trapping of contaminants in the face gap, the wavy face does not suffer this phenomenon.

C603/038/2003 © With Author 2003

Figure 5 illustrates that fluid entering the valley portion of the wave has three paths that it can take during dynamic operation.

Fluid travelling towards the wave peak is compressed and localized pressure increase over the sealed pressure is created. A small portion of this fluid exits across the seal dam as leakage, another portion continues over the wave peak as hydrodynamic load support, the third (largest) portion is recirculated back into the seal chamber, taking any contaminants away from the interface.

This shape has the advantages of bi-directionality, high fluid film stiffness, long life and high reliability.

The laser machining process has evolved to provide the smoothness required providing suitable topographical surfaces for good mechanical seal faces and performance, as can be seen in figure 6.
Seal designs for non-contact operation work with a film thickness greater than approx. 0.5 microns, based upon experiments in water employing hard face (silicon carbide) combinations. Gas seals require a thicker film gap of greater than approx. 2 microns, due to the lower film stiffness in gases.
As leak rate is roughly proportional to the cube of film thickness, seals are designed to run with a very small gap. (For comparison a single sheet of copy paper is approx. 100 microns thick!).

PRACTICAL EXAMPLES

Development of this technology was originally undertaken to provide reliable non-contacting gas seals. This work has been extended to be used in high duty fluid seals, where the operation of conventional flat faces can be problematic, for example where wide variations in the condition of the fluid to be sealed may be encountered.

A few examples of seal designs tested and in operation are included: -

Back-up seal
Figure 7, shows a back up seal designed for gas operation but having fluid retention capabilities. It can be used to support single seals on hydrocarbon services where both volatile and non-volatile primary seal leakage may be expected. It has been developed to provide more reliable secondary sealing in gaseous leakage situations but also allow controlled disposal of any fluid leakage from the main seal.
Leakage values from tests on this type of seal are shown in this Figure 8.
The seal has leakage of well below 1000 ppm of Propane gas at up to 0.17 MPa,
Liquid leakages of up to 2 ml/min have been experience on Diesel at 2.07 MPa.
Many hundreds of these seals are now in successful operation throughout the world, providing safe operation without the complexity of systems required to support dual seal arrangements.

Boiler feed pump seal

A boiler feed pump seal (shown in a dual configuration for testing) is shown in Figure 9. The seal on the LHS is the so-called plug seal to enable pressurisation of the seal on the RHS, which is the seal under study.

A variety of testing conditions were undertaken for this type of design, pressures up to 19.3 MPa, at water temperatures from 49 °C through 71 and 82 °C up to 205 °C, speeds varied from 1200 rev/min to 5550 rev/min.

A seal with shaft diameter of 159 mm is shown, the seal face materials were Silicon Carbide against itself.

The seal was tested with normal water and also with significant air ingestion and abrasive inclusion.

It will be appreciated that the severity of some of these operating parameters, when combined, would have provided an extremely difficult problem with standard faces.

Figure 10 shows the condition of the stationary face after completing a 500 hour test at a speed of 5550 rev/min, with P = 4.03 MPa and T = 65 °C The seal dam and a 'wave' can be seen in excellent condition. Also visible on the atmospheric side of the face at a step below the level of the wear nose is calcium carbonate scale. This indicates the severity of the seal operating condition and re-emphasises the contamination rejection properties of the face design.

Pipeline seal

The seal shown in Figure 11 is being used on pipeline applications.

One such is for Crude Oil/Ethane batches, inclusive significant abrasive content,
sp. gravities from 0.85 to 0.4 and pressure from 2.76 MPa to 7 MPa.

The primary seal faces are Silicon Carbide with waves on the stationary face. The back up seal is of the type shown earlier (figure 8).

Earlier this year the seals had operated for 1100 hours with no reported problems.

This same type of seal is also in operation on another pipeline handling Crude Oil/NGL batches, with frequent stop/starts on pressures from 0.7 MPa to 9.66 MPa.

Satisfactory operation in excess of 16 months had been completed earlier this year.

Once again consideration of the wide variation in operating parameters on these duties shows the difficulties that would have been experienced when using conventionally faced seals.

Other faces

Many seal designs have now been made with laser machined wavy face technology and figure 12 shows a variety of some of the faces that have been produced.

HYDRODYNAMIC SURFACE TENSION TECHNOLOGY

Finally this technology is being adapted to include other load support and leakage restriction mechanisms.

For certain applications the use of surface tension in fluid plus hydrodynamic support is being combined in the so-called Hydrodynamic Surface Tension Technology (HST).

C603/038/2003 © With Author 2003

An enlarged view of the grooves applied near the outside of the seal face together with the waves towards the inner diameter is shown in figure 13 (greatly enlarged).

The waves generate lift for non-contact operation, the grooves provide surface tension bands, which support pressure, drop and eliminate leakage.

The actual face can be seen in figure 14.

This seal is applied to the output shaft of a gearbox on a high-speed pump. Current installations are in excess of 16 in number.

The potential for this modified face topography to provide low energy consumption, low wear; highly reliable seals under varying arduous conditions appear vast.

REFERENCES

1   Salant, R.F. and Key, W.E. 1984 "Development of an Analytical Model for Use in Mechanical Seal Design" In proceedings of 10th International Conference on Fluid Sealing, UK BHRA Fluid Engineering

2   Lebeck, A.O. *Principles and Design of Mechanical Face Seals* , New York; Wiley-Interscience, 1991

3   Iny, E.H. "The design of Hydrodynamically Lubricated Seals with Predictable Operating Characteristics", Proceedings of 5th International Conference on Fluid sealing, BHRA, 1971

4   Young, L.A., Burroughs, J.N., Huebner, M.B., "Use of Metal Bellows Non-Contacting Seals and Guidelines for Steam Applications", Proceedings of the 28th Turbomachinery Laboratory, Texas A&M University, College, 1999.

5   Young, L.A. & Lebeck, A.O., "The Design and Testing of Moving-Wave Mechanical face Seals under variable Operating Conditions in Water", *Lub.Eng.42*, 11,1986.

6   Young L.A., "The Design and Testing of a Wavy-Tilt-Dam Mechanical Face Seal", *Lub.Eng.45*, 5, 1989.

7   Young, L.A., Key, W.E., & Grace, R.L., "Development of a Non-Contacting Seal for Gas/Liquid Applications using Wavy Face technology", *Proceedings of the Thirteenth International Pump Users Symposium*.Turbomachinery Laboratory, Texas A&M University, 1996

8   Young, L.A. & Huebner, M.B., "The use of Wavy Face Technology in Various Gas Seal Applications", *Proceedings of the Fifteenth International Pump Users Symposium*, Turbomachinery Laboratory, Texas A&M University, 1998.

Deformations within a seal

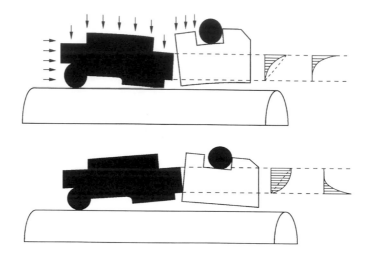

Figure 1

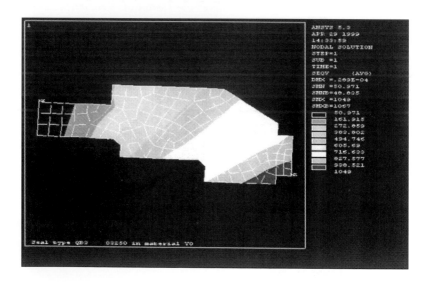

Figure 2

 C603/038/2003 © With Author 2003

Radially Parallel Waviness

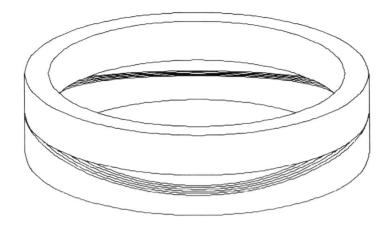

Figure 3

Waviness Features

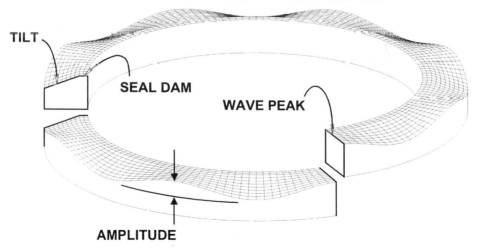

TILT

SEAL DAM

WAVE PEAK

AMPLITUDE

Figure 4

Contamination Resistance
Circulation Effect of Waves

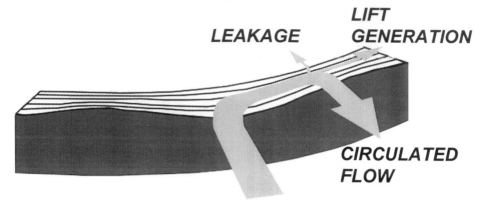

LEAKAGE

LIFT
GENERATION

CIRCULATED
FLOW

Figure 5

Typical Wavy Face

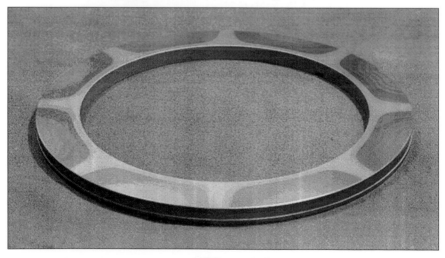

Figure 6

     C603/038/2003 © With Author 2003

BACK UP SEAL

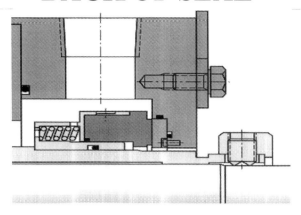

Figure 7

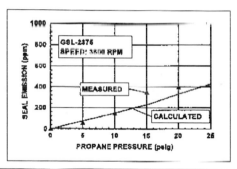

Table 1. Leakage - Wet Conditions				
Seal Size	Pressure (psig/barg)	Speed (rpm)	Fluid	Leakage (cc/min)
2375	300/20.7	3600	Diesel	2
	300/20.7	3600	ISO 32 Oil	2
	600/41.4	3600	Aeroshell 31	3

Figure 8

Boiler Feed Seal – Test Set Up

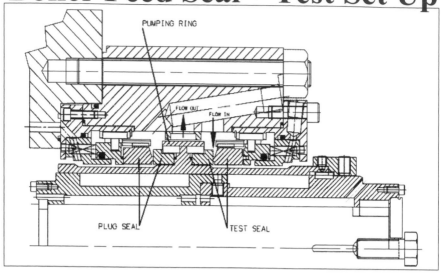

Figure 9

Boiler Feed Test Face – Post 500 hr. Run

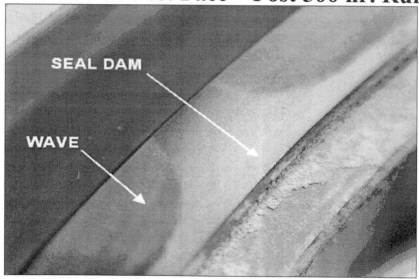

Figure 10

Pipeline Applications
HDHW 4500 / GSL 4437

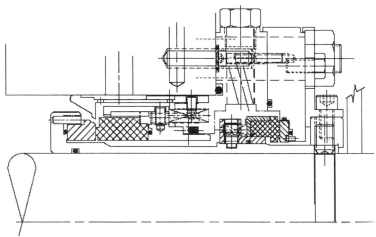

Figure 11

Variety of Laser Machined Faces

Figure 12

Hydrodynamic Surface Tension (HST) Technology

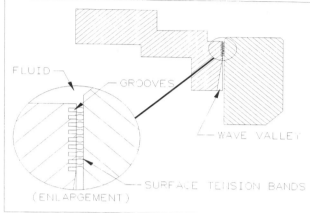

Figure 13

High Speed Pump – G/Box Seal

Figure 14

Authors' Index